JACOB'S LEGACY

JACOB'S LEGACY

A Genetic View of Jewish History

David B. Goldstein

Yale University Press / New Haven & London

Published with assistance from the Louis Stern Memorial Fund.

Designed by Nancy Ovedovitz and set in ITC Galliard type
by Duke & Company, Devon, Pennsylvania.
Printed in the United States of America.

Library of Congress Cataloging-in-Publication Data
Goldstein, D. B. (David B.)
Jacob's legacy : a genetic view of Jewish history /
David B. Goldstein.
p. ; cm.
Includes bibliographical references and index.
ISBN 978-0-300-12583-2 (cloth : alk. paper)
1. Jews—History. 2. Jews—Origin. 3. Human genetics. I. Title.
[DNLM: 1. Jews—history. 2. Anthropology, Physical. 3. Genetic
Predisposition to Disease. 4. Genetics, Population. 5. Jews—
ethnology. 6. Jews—genetics. WZ 80.5.J3 G624j 2008]
GN547.G65 2008
909'.04924—dc22   2007043583

A catalogue record for this book is available from
the British Library.

The paper in this book meets the guidelines for permanence
and durability of the Committee on Production Guidelines
for Book Longevity of the Council on Library Resources.

10 9 8 7 6 5 4 3 2 1

This book is dedicated to the memory
of my grandmother, Estelle Goldstein, who
somehow tied up my cultural moorings
while no one was looking

.   .   .

CONTENTS

CONTENTS

My entry into Jewish genetic history was both personal and professional.

For a long time, when asked how I came to study the genetic history of the Jews, I had ready rather pat answers about the intersection of population genetics, human evolution, and the oral and written traditions of Jewish civilizations. Those subjects have fascinated me for twenty years and remain at the heart of my intellectual life. But with bits and pieces of this book in front of me that recapitulate the better part of a decade's worth of work, it is also clear that my evolution as a genetic historian is more personal than simply the interaction of genetics, evolution, and history. This is for me in part a personal story.

The first turning point occurred during my graduate career in the Stanford University lab of the great population geneticist Marcus Feldman. I was trying to develop abstract mathematical models that might support general principles about the evolutionary process. In particular, I was studying the evolution of ploidy, that is, the number of sets of chromosomes a species has. Humans (and almost all animals) have two—they get one set from Mom and one from Dad. In other words, we are "diploid," whereas bacteria are "haploid" (having just one set); as for plants, you never know how many sets of chromosomes they are going to have. My task was to figure out whether inheriting different versions of the same gene from each parent in a beneficial combination ("overdominance") favored the evolution of diploid organisms. Said another way, the question I addressed was whether it is formally possible that the reason for diploidy is that sometimes having two different versions of a gene is better than having only one version. For some months I struggled with different algebraic formulations to identify the exact conditions under which that might happen. When I use mathematics it is usually only to prove something I already suspect, and this case was no exception. Even so, it was a long process. When I finally arrived at what I thought was the answer, I took it to Marc's office. While I waited by his desk, he looked over my work with his usual meticulousness (successful population geneticists are many things, but casual is certainly not one of them). After no more than twenty minutes and perhaps half a page of algebra he nodded and said simply, "It's correct." There was, apparently, a serious shortcut to the answer that was quite obvious (to Marc, at any rate).

Although I was certainly glad to have applied the algebra correctly, the gratification was fleeting. I quickly realized a couple of things. First, I would never, under any imaginable circumstances, turn into the mathematician that Marc Feldman is. Second, I was not really a theoretical population geneticist by inclination. The problem with theoretical population genetics is that it is often about trying to model what *could* happen. I was, and remain, much more interested in what actually *did* happen. And so I began to turn my attention to human evolution.

Right away I had a huge advantage: the greatest human evolutionist of my time worked just across the street. He was Luca Cavalli-Sforza, the Italian geneticist who has towered over the field since the 1960s. I was in awe of his accomplishments, therefore, when I first made my way to his lab sometime in 1993, but it immediately became clear that he was not just a legend but very much an active intellectual force in the field. He was in the process of developing a powerful new kind of genetic marker for use in human evolution, called microsatellites, which came to be an essential part of my own work. In the early 1990s, Luca's lab began conducting experiments that would eventually show that microsatellite variation was sufficient to assign individuals accurately to their historical continent of origin. Not too much was made of this discovery at the time, although some ten years later, in 2004, Luca and Marc published an extension of these studies that finally received the recognition that they deserved.

When I went to see Luca, I had been working on a somewhat arcane piece of genetic modeling. He asked me what I was trying to do and why, and what was the most interesting thing that could possibly come of it. I realized that the most interesting thing wasn't really all that interesting to me.

And so here was the next turning point in my career. What Luca made clear to me is something I clarify for all of the students and postdoctoral fellows I mentor: if the most interesting thing that might happen as a result of your work still doesn't excite you that much, forget it—if you don't really care, why should anyone else? But if you do care, that is really all you need (along with a bit of luck here and there). Even if things don't turn out as expected, something will probably come of your efforts anyway. Better to run the risk of spectacular failure than face the certainty of accomplishing something dull.

Luca suggested that if we could figure out a good way to use microsatellites to understand how long ago human populations separated from one another, we would be onto something. Luca, Marc, and I, in various combinations, wound up publishing eight papers on the subject, and I began to make my reputation in the field, although in the end things did not work out exactly as we had hoped (they rarely do in science). The rate at which microsatellites change ("mutate") was too poorly understood to give us the accurate information we needed (and, along with the rest of the field, we came to realize that most human populations didn't really "separate" at specific time points). Nevertheless, I learned a great deal from the endeavor, gaining the experience that would allow me to figure out a new way of determining the age of ancient Jewish chromosomes.

I suspect that had I no Jewish heritage, this work would likely have never led me into Jewish genetic history. Events, however, primed me to look for ways to translate my professional activity into some kind of connection to my own ethnic background. I remember, for example, feeling agitation and anger during the first Gulf War when Israel was being bombed by Iraq and America was instructing the Israelis to sit tight. My first, and I suppose somewhat childish, impulse was to enlist in the Israel Defense Forces. I had just started graduate school at Stanford, though, and was afraid I might lose my only chance to pursue a serious career in genetics. Still, throughout the Desert Storm period of 1990–1991, I felt guilty for not doing anything serious and angry with everyone around me who didn't seem to care in the least.

To compensate for the guilt of having chosen my career over Israel (not to presume that Israel ever had any need of me), I began to study Hebrew. Having never been bar mizvahed, I didn't recognize a single letter of that crazy ancient script, but with a pile of books and tapes courtesy of a friendly undergraduate student doing part-time work in the library, I set aside a couple of hours each evening to learn to read and write from right to left, make sense of the alphabet, pronounce my guttural "ch" sounds with sufficient oomph, and appreciate the hard and logical structure of the language. A few years later I married the friendly undergraduate at a nearby synagogue. In one of history's little quirks, I have Saddam Hussein to thank for both my marriage and the fact that I can now read and write Hebrew.

My fellow graduate students Aviv Bergman and Sharoni Shafir kept me stocked with Israeli pop music, which I would translate crudely and sing horribly in the shower. In the 1980s and 1990s, Israeli rock was transcendent, somehow both powerful and starkly beautiful. I began a lifelong love affair with it, especially the music of singer-songwriter Yehuda Poliker. A few years after Desert Storm, I had the pleasure of seeing Yehuda in concert at the Roman amphitheater in Caesarea on the Mediterranean coast. The memory of that evening says as much about why I wanted to do the work described in this book as anything else might. I was particularly struck by modern Hebrew rock lyrics blaring through massive loudspeakers in an amphitheater where two thousand years earlier Jews must have been tortured by the Romans. Now here I was listening to a young, out-of-the-closet gay Israeli belting out a song about how the Holocaust remains with him (he is the son of survivors from Greece). Surrounded by rowdy young Israelis boisterously singing along, I felt part of it all, even able somehow to understand the lyrics. The connection with that community that night, hard to define but tangible, told me something about how the Jews had survived all that they had and somehow remained a people. I often think back to that concert and those kids taking off their shirts and swirling them around, those Cohen Y chromosomes and varied mitochondria that may have started there and somehow found their way back after two millennia. For me, that is still a major part of what this is all about: the improbable magic of it all.

In addition to Yehuda Poliker, I owe a debt to Shalom Chanoch, an Israeli Bob Dylan who often chastises his countrymen for their arrogance and intransigence but who nonetheless writes that upon "sitting in San Francisco on the water" he cannot wait to get back to his little country "hot and wonderful." I stayed in graduate school and stayed out of the Israel Defense Forces, even though I never quite gave up the idea of enlisting. Even now, in my early forties, when I am pushing a little hard on a run, I tell myself to keep going—with the world as crazy as it is, who knows? All of this gave me an intense interest in using my slowly growing professional reputation to some-how establish a strong link to the Jewish world and to Jewish history. But I found no easy way in until I received an auspicious e-mail.

I had just been recruited to Oxford University. It was my first job, and I wasn't sure which way was up. Providentially, Linda Partridge, a senior evolutionary geneticist at University College London, took me under her wing (perhaps at the urging of my chair at Oxford, Paul Harvey). Even before I arrived in England, Linda was writing me with advice about how to navigate the British science scene. And one of the things she did was to tell Neil Bradman to drop me a note.

Neil and I hit it off instantly. Quite simply, Neil went for the jugular every time: did the priests really inherit their priestly status, or did some just decide to be priests? Were the Lemba really descended from the Jews? And even bigger questions haunted—and continue to haunt—him: Are Ashkenazi Jews the remnants of the Khazari Empire? I was hooked. Neil became a student, collaborator, and friend. Over the years we've marched through quite a few questions and, to my amazement, even gotten a few solid answers. The work I discuss in this book derives in large part from his enthusiasm, competence, and unflagging confidence.

The last catalyst for this book came years later when I was describing our lab's work at a synagogue in Burlingame, California. My dad, who had come along, had warned me beforehand that what I needed to do was to describe the science in a way that nonscientists could understand. His advice made all the difference. After I spoke, it seemed like the entire audience peppered me with questions—they understood it all, I remember thinking. We talked about the Jewish priests, the meaning of the Y chromosome lineage, and who it might have come from. One of the women in the audience pointed out that the story was really about the wives, not the priests: if there had been even a small amount of funny business going on over the years, there could be no continuity of the priestly Y chromosome line. I had simply never thought about that before. Finally, another woman told me that I really needed to write a book about all this stuff because people wanted to know about it.

While of course not unique in their interest in genetic history, the Jews have tended to emphasize ancestral connections. The centrality of ancestry in Jewish identity is perhaps best illustrated by the declaration, repeated by Jews each year in the Passover service, that "we" were slaves of Pharaoh in Egypt. Dispersed throughout the world, Jews today see themselves as a continuation of the ancient people who fled Egypt for the Promised Land in Canaan. This book describes a few of the small pieces of their story visible to an interested geneticist.

ACKNOWLEDGMENTS

First and foremost I must express my debt to Neil Bradman. As will become clear within the next few pages, not only would this book not exist without him, neither would the body of work that it describes, or at least not my involvement in it.

Science has increasingly become a group activity, and many other colleagues and friends contributed indispensably to this work. In addition to Neil, I must draw particular attention to the efforts and contributions of Mark Thomas and Mike Weale. As for the many others who contributed, their names and roles are revealed in the pages that follow.

I offer my sincere gratitude to my friend and colleague Misha Angrist for taking an undisciplined and unruly early version of this book and editing it into structural coherence and consistency. He taught me how to write more accessibly, less arrogantly, and—well—better. After Misha I was blessed with two gifted editors at Yale University Press, who organized a useful and accurate set of reviews and greatly improved the manuscript with suggestions small and larger. Finally, I express my deep gratitude (and amazement) to my agent, Georges Borchardt, for sticking with the project as the months stretched into years. Without his never-ending patience and support this book would not have appeared.

JACOB'S LEGACY

INTRODUCTION

Three thousand years ago a small kingdom emerged in the southeast corner of the Mediterranean, wedged between the remnants of two great empires (figure 1). That kingdom, the United Monarchy of the Hebrews (or Israelite Kingdom), lasted only seventy-five years, the span of a single human life. But during that brief period and the years leading up to it, the Hebrews established a national identity strong enough to carry them through repeated episodes of exile and subjugation and, in one of the most dramatic political movements of modern times, back to their original land. Judaism and the Jewish people, however, are only the most direct legacies of the ancient Israelites. This rather insignificant group living on the fringes of settled society, whom the Egyptians may have called "Habiru," developed moral, philosophical, literary, and religious traditions of such novelty and power that they eventually came to wield remarkable influence on the civilizations of the Near East, Europe, and the world.

Who were the ancient Hebrews? Where did they come from? What ultimately became of them? Most of our information comes from the oral traditions and written accounts of the Jews themselves, and surprisingly little is known with any certainty. We know the language

1

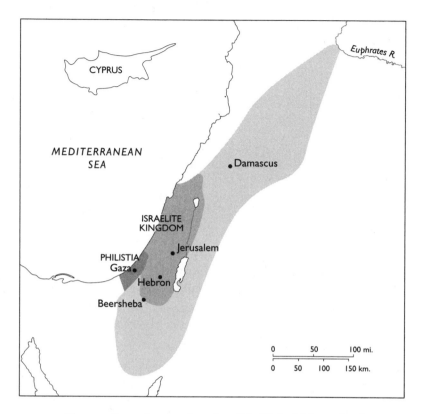

Figure 1. A rough approximation of the United Monarchy of the Hebrews, or Israelite Kingdom, under King Solomon's reign (ca. tenth century B.C.E.). Map by Bill Nelson.

and customs of the Hebrews were similar to those of other Semitic peoples. The Hebrews, or some group within them, may have had a nomadic lifestyle before inhabiting the cities of Canaan. But how much time they may have spent in Egypt, the extent to which they inter-married with non-Semitic peoples in Egypt or elsewhere, how far any particular Jewish population traces back to the early dispersions of the

Hebrews—all are matters of speculation. Over time, the Diasporan communities coalesced into the Ashkenazi, Sephardi, and Mizrachi (Oriental) traditions of modern Judaism. Most scholars agree that the Jews of the modern world have their origins, at least in part, in the Hebrews of the Near East, although some probably derive from the conversion of non-Semitic peoples. But how much does the continuation of the Jewish people and of Jewish custom owe to a genetic continuity of the Jews and how much to cultural forces operating independently of hereditary connections? Without genetics this question remains simply unanswerable.

This book describes a few specific aspects of Jewish history that have been studied using the genetic legacies carried by living people that call themselves Jews. Genetics can never, however, replace, or even really compete with, the painstaking work of archaeology, philology, linguistics, paleobotany, and the many other disciplines that have helped resurrect some of the lost stories of human history. No new genetic technologies or careful mining of the genetic heritage of any people will ever unlock stories like those that flowed from the Rosetta stone, which revealed pieces of Egyptian history that would have been forever lost if hieroglyphics had remained opaque to us.

But genetics is slowly earning a place in the historical sciences. Our narratives describing the histories of peoples and events, from the Aryan invaders of India to the Viking attacks on the British Isles, are all being augmented and refined by genetic analyses in a field now often called genetic history.

Genetic history is both more and less significant than it is depicted in popular accounts. It is less significant because the historical insights that can be achieved with genetics are always very specific and often fairly modest. Caught up as it is in the excitement of modern science, genetic history's power and importance are often overstated, whereas the real grandeur and detail of human history can only be seen in the context of our archeological and written legacies and, of course, our memories. But at times genetic history stretches the boundaries of its scientific formalism and hints at answers to bigger questions: What makes a people a people? What binds them together through time? What alienates them from some and aligns them to others?

To appreciate fully what genetics does and does not tell us about our history requires an understanding of inheritance and the genetics of populations. This book reflects my conviction that nonscientists can understand the principles of both genetics and genetic history and that the history of Jewish populations represents a powerful and fascinating model for illuminating these principles. In the chapters that follow, I have attempted to explain the science from first principles and with minimal disruption of the narrative flow. I use specialist terminology only when necessary, and I always provide definitions in everyday language. In addition to the necessary genetics, I also incorporate the history of the Jews to provide the historical context within which the work has been undertaken.

I begin with the biblical and oral traditions surrounding the Jewish priesthood and priestly tribe. Here I explore whether these ancient traditions have a clear genetic legacy among contemporary Jewish priests, the Cohanim. To address this question, I use patterns of genetic variation on the human Y chromosome, which records the history of the males of a population. From the Jewish priests, we move on, in chapter 2, to the exiles. Our first stop is southernmost Africa, where a small group of Bantu-speaking people, the Lemba, claim a three-thousand-year-old origin in ancient Israel. Here again we use the Y chromosome to search for signatures of a Jewish genetic heritage of the Lemba. From this surprising outpost of Jewish identification, we shift our focus, in chapter 3, to the whole of the Jewish Diaspora, beginning with the question of whether the genetic legacy of the priestly tribe of Levi points in directions distinct from that of the more elite priestly class, the Cohanim. Up to this point our focus is the part of our genome that is influenced only by males—their movements and their particular demographic histories. These analyses are blind to female history.

Chapter 4 expands to parts of the genome that record and reflect female history. Here we delve into the question of how much the maternal genetic heritage of Jewish populations differs from the paternal heritage and if it is even of Middle Eastern provenance at all. At the heart of this study is the question of whether the differences between maternal and paternal lineages are influenced by the long-standing custom that Jewish identity descends through the mother.

In chapter 5, I address the medical ramifications of our increasing knowledge of genetic variation in Jewish populations, focusing particular attention on areas of overlap with Jewish genetic history. A number of genetic diseases (or particular mutations in disease-causing genes) have clustered within Jewish populations, especially the Ashkenazim. I discuss the historical and genetic bases for this and even consider a radical hypothesis or two as to why history may have played out the way it did, including one advanced by Henry Harpending and his colleagues that Jews were selected for increased intelligence. This hypothesis, unsurprisingly, is popular among some American Jews.

Chapter 5 also attempts to demonstrate how the consequences of human genetic history continue to resonate in the practice of modern medicine. As our ability to identify, diagnose, and treat diseases with genetic components increases, we will be confronted with a growing list of ethical dilemmas accompanying our bounty from the tree of genetic knowledge. A debate in medical genetics that aired in newspapers in 2005 throughout the world exemplifies the dilemma. In June of that year the FDA approved BiDil, a drug for congestive heart failure, specifically for use by African American patients—the first so-called racial drug. Many academics, notably but not exclusively social scientists, attacked the rationale behind this decision as a dangerous perpetuation of outdated concepts of racial differences. I explore that rationale and contemplate what it might mean for the future of medical practice as we learn more and more about genetic variation among different groups. The subject has obvious consequences for Jews in terms of identity and health.

Finally, chapter 6 describes some of the startling technological advances that are allowing researchers to assess variation comprehensively throughout the whole genome and how these advances may be used in genetic history.

Although the stories told here cannot alone establish any new paradigms in our understanding of Jewish history, they do illuminate the complexity of that history. From a demographic and genetic perspective we see the history of the Jews as an intricate combination of forces, both genetic and cultural, woven together and operating with different strengths at different times. No single event or connection unites the Jews; different strands connect different groups at different times to give us what we now call the Jewish population. Before launching into the genetic stories there is just one last question to address: Why the Jews?

As far as we can tell, genetic history has fascinated all peoples at all times. Just outside the old city of Valletta, the capital of Malta, stands the country's premier hotel, the Phoenician Meridian. Ask the Maltese cabdriver taking you there, and he is likely to tell you that the Maltese are an ancient people descended from the Phoenicians, Levantine mariners and close relatives of the ancient Israelites. The Phoenicians established trading posts throughout the Mediterranean, including on Malta, and competed against the Greeks. From their North African colony, Carthage (just a few miles from Tunis), they conducted an epic but doomed struggle against an emerging Rome during the third and second centuries B.C.E. (Before the Common Era, equivalent to B.C.). It was the Phoenicians, too, of course, who gave us Phoenician, the world's first fully phonetic script. Similar stories could be told about Hungarians, Icelanders, Japanese, Finns, Greeks, the Parsi of India, or any number of people. The Jews are by no means unique or even unusual in their interest in their history and in the role that genetics might play in uncovering that history. There are books to be written about the genetic histories of peoples throughout the world. This is but one set of stories that could be told with genetics.

CHAPTER 1

Keeping God's House: Y Chromosomes and Old Testament Priests

> They shall teach Your statutes to Jacob,
> Your law to Israel; they shall place
> incense before You, and a whole burnt
> offering on Your altar.
>
> *Deuteronomy 33:10*

When people learn that I use genetics to assess relationships among people or groups of people, they are often puzzled by an apparent contradiction. A segment of the academic community has for years insisted that we are "all the same," that there are no meaningful genetic differences among populations and no biological basis for race and ethnicity. If that is so, how can we use genetics to look at the relationships among groups of people? Some of this thinking traces back to decades-old work by the eminent evolutionary biologist Richard Lewontin, who showed that 85 to 90 percent of the total genetic variation humans harbor is due to differences among individuals within a given group (racial, ethnic, or otherwise). Only about 10 to 15 percent is due to average differences between members of any two different groups.

Think about a simple trait such as height. If one were to look at a first-grade class and a high school graduating class, each of, say, thirty kids, there would be a great deal of variation in height among the students. Most of the variation, however, would be explainable and predictable by what class any given student was in—either the first grade or the twelfth grade. On the other hand, if one compared first graders with second graders or two classes of the same grade, there would still be variation in height, but most of the variation would be due to differences among individuals *within* classes, not to differences on average *between* the classes. This is how it is with human variation. Looked at another way, if the whole human race were destroyed except for, say, one Swedish village, that small group of survivors (or, in genetic terms, founders) would nonetheless preserve the majority of our genetic variation. This apportionment of our total genetic variation is now broadly accepted. Most of our genetic variation emerges from differences among individuals, as opposed to groups, because we are demographically a young species: most human groups carry with them the variation that was present in our African ancestors, from whom we all arose some hundred thousand years ago.

For most of the past twenty-five years or so, the scientific community has followed Lewontin in arguing that a 10 percent genetic difference among members of different groups amounts to a small amount of genetic variation. Social scientists in particular have shown a willingness to extrapolate from Lewontin's study that race and ethnicity are biologically meaningless, hence the refrain that "we are all the same." The data I present here, however, showing clear genetic differences among the members of population after population, as well as hundreds of other published studies on the genetics of disease, drug response, and human variation, argue otherwise. Nor should we be surprised: there are ten million sites in the human genome that show differences among individuals (even among relatively small groups of individuals). If only a small fraction of these showed differences that correlated strongly with race or ethnicity (or, in less loaded terms, with geographic ancestry), that would still include hundreds of thousands of genetic differences. Put another way, the Swedish village we imagined would retain most of the variation that we have, but not all of it. And more than that, we would be able to predict many, many variations that would likely be lost because they are known not to be common in Sweden. Saying that most of the variation is common in the human genome still leaves a lot of room for differences.

Not long ago genomicists Steve Scherer and Matthew Hurles and colleagues from around the world published a paper in *Nature* that shocked even those of us entrenched in the "we are not all the same" camp. They reported that, although humans all have more or less the same DNA sequences (the ten million differences I alluded to notwithstanding), they can vary substantially in the *number of copies* they have of any given stretch of DNA. Thus, any two individuals will likely exhibit copy-number variation over 12 percent of the genome on average. Bottom line: the human genome is both complicated and very big. Although most of the variation is due to differences between individuals, that still leaves plenty of room for genetic differences that inform us about the histories of groups of people.

The group of people I became interested in was the Jews, and as I say in the preface, it was Neil Bradman and his e-mail that spurred my pursuit. Neil was a prospective graduate student, but, as I was to learn, he was far from typical. Twenty years my senior, he had been a businessman for most of his adult life. For someone like me, sequestered in the academic world, he may as well have been writing from another planet. Fortunately for me, he spoke my language. Having succeeded brilliantly in business, he'd turned back to science, having some years earlier earned a master's in parasitology. He persuaded the biology department at University College London to establish a research collaboration, and making use of his early training, he kicked it off with nematodes. The nematode, or roundworm, is a simple organism with both free-living and parasitic forms that has earned a permanent place in science. The adult has only 959 somatic cells, and its simplicity and its transparent cuticle make it possible to follow the development of individual cells, from their beginnings to their eventual roles in the soma (or body)—in other words, to establish an entire "fate map" of all the cells of the nematode. In 2002, Sydney Brenner, Sir John Sulston, and H. Robert Horvitz shared the Nobel Prize in Medicine for their work in developing the nematode as a model system for assessing how genes influence development. But the study of nematodes was not where Neil would make his name. After a false start with the roundworm, Neil moved permanently into genetic anthropology.

Evidently he and Karl Skorecki, a clinical nephrologist at the Rappaport Faculty of Medicine at Haifa's Technion Institute, understood at about the same time that the oral tradition of patrilineal inheritance surrounding the Cohanim, or Jewish priesthood, could be easily and directly tested simply by looking at genetic variation among the Y chromosomes of Jews who do and do not claim priestly descent.[1] As I discuss in greater length below, the Jewish priesthood is an ancient tradition, and by custom a Jewish high priest, or Cohen, is an individual whose father was a priest (*Cohen* is the Hebrew word for priest, and the priesthood is totally independent of the much more recent Jewish tradition of rabbis). Neil's insight was to realize that if the status of priesthood really was inherited from father to son there should be a record of it in the genomes of people today who consider themselves priests.

Neil sent his son to the beach in Tel Aviv to collect DNA samples, and Karl enlisted Mike Hammer, an eminent genetic anthropologist at the University of Arizona renowned for his work on the human Y chromosome. The team looked at one of the workhorses of Y chromosome genetics—an insertion of a chunk of DNA called an Alu repeat. They also looked at a microsatellite marker, a highly variable stretch of DNA sequence about which I'll say more later. They compared the forms of these markers between priests and nonpriests in both Sephardi and Ashkenazi communities. The team hit a home run their first time at bat. They found a striking difference in the Y chromosomes of priests and nonpriests, a difference that was consistent in both the Ashkenazi and the Sephardi communities. They published their results in perhaps the most venerated scientific journal of all—*Nature,* the journal that carried Watson and Crick's proposed structure for DNA and a huge number of other distinguished discoveries.

But Neil wasn't satisfied. He wanted to expand his work, greatly increasing the amount of genetic variation he studied, collecting more samples, and analyzing more of the highly variable microsatellites. He had the good sense to hire Mark Thomas, a postdoctoral fellow from Cambridge University. Nothing enters or leaves the Thomas lab without a complete recording of time, place of origin, and contents; freezers are accessed only at particular times, laboratory buffers can be made up only under his supervision, and on and on—every "i" dotted and every "t" crossed. Beyond meticulous, Thomas is a scientist's scientist, the model of choice when the press needs an action shot. He occasionally makes those of us with a more California style a little crazy: when you ask for a sample for an experiment you may well be told that the freezer in question is not accessed on Tuesdays, only on Thursdays. You laugh, but Mark is serious. Yet I would not have traded his meticulousness for all the California Zen on offer. Once, when someone accused our group of sample manipulation, I had not a second's concern. In this collaboration the samples had all been received, stored, and inventoried by Mark. If anyone on the planet could document precisely what happened to each and every sample he had ever handled, it was Mark Thomas. And so he did. The accusation was summarily dismissed by officials at University College London.

The laboratory side well in hand, Neil wanted statistical help, too. My friend and mentor the biologist Linda Partridge directed him to me. By that time, after the extensive work I'd done with Luca Cavalli-Sforza, I had earned the moniker Mr. Microsatellite and was an imminent transplant to Britain, nervously starting my first job at beguiling and frustrating Oxford University. In his e-mail, Neil introduced himself to me, told me what he was doing, and extended a cordial invitation to Shabbat dinner at its home. He signed himself "Neil Bradman (of the Jews)." I had no way of knowing it then, but Neil Bradman of the Jews was to become one of my closest research partners over the next five years.

At the time, I thought either this guy, like me, had serious cut-and-paste issues in his e-mails or, more likely, he was barking mad. A bit wary, I declined the dinner invitation, although I did agree to meet up with him. That remains one of the best professional decisions of my career. I have since had the great pleasure of attending more than one of those dinners, where once the prayer over the bread is said, Neil will rip off chunks of challah and fire them toward each of his guests in turn. Naturally the newest guest is the first recipient, ideally in the forehead. Hours of conversational free-for-all follow, with Neil basking in the relaxed, congenial atmosphere he has created.

The concept of "Jewish priests" is both alien and exotic to most contemporary Jews, myself included before Neil's introducing me to it. Outside of a few Orthodox sects led by charismatic rebbes, the idea of a central religious figure directing the spiritual, cultural, and political lives of his Jewish community is indeed a foreign one. But this was exactly the role of Old Testament priests. According to the book of Exodus, they performed their functions in flamboyant costume. A high priest's vestment, or ephod, was a sleeveless purple robe whose hem was adorned with golden bells and pomegranate tassels in dazzling shades of red and purple. On the shoulder piece were two onyx stones engraved with the names of the twelve tribes of Israel. A breastplate was inscribed with twelve different gems, each representing a tribe.

That most of us, even those thought to be descended from the first priest, Aaron, have no frame of reference for Jewish priests is hardly surprising: since the destruction of the Second Temple in 70 C.E., a functional Jewish priesthood has ceased to exist. So, too, have once-central priestly rituals such as animal sacrifice and anointment with sacred oil. But over the thousand years preceding the last gasp of the Jewish revolt against the Romans in the second century C.E., priests exerted a profound influence on Jewish life, both sacred and secular.

In this chapter I attempt to trace the genetic lineage of the Old Testament priests, or Cohanim, from antiquity to the present day. But to understand that lineage in a meaningful way, we must first consider the Cohanim's biblical origins, their evolution (and cultural near extinction), and the significance of the priesthood in ancient Israel. Once we grasp the historical context within which Jewish priests went about their business, we can contemplate how and whether their line might have persisted down through the years and what the genetic evidence is for such a scenario.

Origins of the Cohanim

The Hebrew Bible records the creation of the Israelite priesthood; unfortunately, not much else does. Only much later sources address the subject in any detail. And the Bible itself appears to lend credence to two interpretations of the origins of the priesthood. The first is simply that the priests were synonymous with the tribe of Levi—that is, the tribe of Moses and his brother, Aaron. The alternative view of priestly origins is that, at least initially, genealogy was less important than professional credentials. This version holds that priestly affiliation was not exclusively ancestral; rather, priesthood tended to consolidate within familial and clan lines but remained open to outsiders. These two views have important implications for our efforts to identify a continuous genetic lineage spanning from Old Testament priests to their modern descendants.

In Deuteronomy 33:8, Levi, one of the patriarch Jacob's twelve sons, is described as having divinatory powers. In particular, he manipulates the "Urim and Thummim," an oracular device consisting of two lots or stones that was designed to give a yes or no answer. Perhaps the closest modern equivalents are the Ouija board and the Magic 8-Ball toy.

The Levites, the descendants of Levi, constituted the smallest of the twelve tribes. It had a number of distinctive characteristics. First, it was the only tribe not allotted a fixed territory, and its members were counted separately in the census. Second, in Numbers 3:12, the Levites were designated a priestly tribe, replacing the prior priestly class made up of firstborn Israelite sons.

As the designated priests, the Levites were given a variety of specific responsibilities. After the Exodus from Egypt, they were appointed guardians of the sacred tabernacle (the Tent of Meeting) and carriers of the Ark of the Covenant. Biblical scholars have learned a number of other features of the Levites from the amalgam of Pentateuchal priestly materials that is thought to reflect the voice or at least the views of the priesthood (usually therefore referred to simply as "P"): the need for tithes to support the Levites (since they lacked land of their own), obligatory sacrifices they were to make at various festivals, and instructions as to how they were to be treated and what was expected of them by other Jews.

Numbers 3, however, also makes clear that being a Levite was not synonymous with being a full-fledged priest. Only Aaron's descendants were anointed priests. These descendants were the Cohanim, the Hebrew word for priests (or Cohen in the singular). With the construction of the First Temple, one Aaronite family, the Zadokites, became hereditary high priests (Zadok is said to have been a tenth-generation descendant of Aaron who functioned as priest to Kings David and Solomon). Thus, under this scheme, we now have the makings of a three-tiered hierarchy of priestly folks: the tribe of Levi as a whole; the subset of Levites known as the Cohanim, that is, the presumptive male descendants of Aaron; and a subset of the Cohanim themselves, the high priests, or Zadokite individuals selected from the Cohanim who assumed ultimate responsibility for specific religious and administrative duties.

Beyond divination, priests were called on to carry out a range of tasks. At the head of the list was ritual sacrifice. Priests maintained exclusive control over sacrifice and thereby wielded considerable power. Those farmers who were able to deliver choice cuts of sacrificial meat, for example, might be afforded greater status in the eyes of the ruling priests. In addition, Leviticus 6:14–15 describes priests providing meal offerings for the whole body of the priesthood. In all likelihood, priests were also intimately involved in temple-based rituals marking the Sabbath, the New Moon, and other holidays and festivals. Priests also served as lawgivers, legal scholars, and, more practically, magistrates charged with adjudicating criminal and civil disputes. These roles implied literacy, which again would have conferred exalted social status.

In the summer of 586 B.C.E., the Babylonians burned the First Temple, effectively ending worship and sacrifice in Jerusalem for decades. In 2 Kings 25:18, we learn of the executions of the chief priest Seraiah as well as his second in command and several other priestly "keepers of the threshold." We know from the book of Ezra that a number of surviving priests wound up in Babylon during the fifty-year exile. From ancient papyri and from the writings of first-century Jewish historian (and, to some, Roman apologist and even traitor) Flavius Josephus, we know that Jewish temples existed in Babylon during the exile. Some historians have speculated that these temples were the predecessors of modern synagogues, local places for communal worship that became a hallmark of Diasporan Judaism in the Common Era.

In 539 B.C.E., the Babylonian Empire fell to the Persians, to the great benefit of the Jews. Just as the aftermath of World War II and the Holocaust would pave the way for the creation of the modern state of Israel in 1948, so too did the end of the Babylonian reign set the stage for the restoration of the Jewish presence in Judea and the construction of the Second Temple. The most tangible stage-setting event occurred in 538 B.C.E. when the Persian King Cyrus established a policy of reconciliation and repatriation of deportees—and their various deities—to their homelands. As recounted in the book of Ezra (1:1–4) and summarized in Joseph Blenkinsopp's *Sage, Priest, Prophet*, Cyrus's proclamation led to the return of some fifty thousand Babylonian Jews, who wasted no time in constructing an altar, offering sacrifices, and celebrating the harvest festival of Sukkot in Jerusalem. A case could quite reasonably be made that without the Persians' taking Babylon and allowing the return of the Jews to their homeland, Jewish cultural continuity would have been lost.

Despite the haste with which Jewish religious practice was restored, the Second Temple would not be completed for another twenty years. Joshua, the first high priest of the Second Temple era, and his colleague the governor of Judah, Zerubbabel ("begotten of Babylon"), were instrumental in reestablishing Jewish worship in the city and overseeing the Temple's construction. In Zechariah 4:14, the men are referred to as "the two anointed who stand by the Lord of the whole earth." This type of combined secular-sacred leadership (governor plus priest) became the model for Jewish governance during the period and reflected growing priestly influence in both prophecy and affairs of state.

In addition to Joshua, no survey of Second Temple priesthood would be complete without a discussion of Ezra, a son (or possibly grandson) of the executed priest Seraiah and a presumed lineal descendant of Aaron. Ezra was a scribe skilled in the law of Moses. In perhaps his most famous scene on the biblical stage, Ezra is said to have read the complete scroll of the Torah aloud to the entire population after Nehemiah (later a governor of Judea) reconstructed the wall surrounding Jerusalem in the mid-fifth century B.C.E.; he and his fellow Levites offered explanations and interpretations as they went. Ezra is also famous for his condemnation of intermarriage and for editing and assembling the canonical pieces of the Old Testament.

The Hellenistic period that followed the construction of the Second Temple (from Alexander's conquest of Judea in 332 B.C.E. to the establishment of the Hasmonean monarchy in 141 B.C.E.) brought profound changes to religious practice among the Jews. As Lee I. Levine recounts in Hershel Shanks's *Ancient Israel: From Abraham to the Roman Destruction of the Temple,* under the Greeks faith and certainty were superseded by doubt, hesitancy, and skepticism. Jews were torn between maintaining their identities and assimilating themselves into Hellenistic culture. For a while, it appeared they could do both. Under the Seleucid King Antiochus III, the priests were officially recognized as community leaders and accorded exalted status in Jerusalem.

Yet some twenty-five years later (about 175 B.C.E.), the Jewish priest Jason is said to have purchased the position of high priest from Antiochus IV along with the right to convert Jerusalem into a Greek polis. Three years later, another priest, Menelaus, imitated Jason and tried to buy the office of high priest for himself, thereby supplanting Jason. Even worse, Menelaus plundered the Temple treasury to come up with the cash for his bribe. As often happens when the leaders of a people turn against one another, those outside the fray showed no mercy. Despite the bribes of Jason and Menelaus, Antiochus massacred, pillaged, and sacked parts of Jerusalem. By 167 B.C.E., he had banned circumcision and religious observance. Even more galling, he forced the Jews to abandon kashrut (dietary laws) and brought idolatrous worship into the Temple itself.

Within a year, in a small town in northwestern Judea, a priest named Mattathias and his five sons had had enough. Mattathias killed a Jew in the act of making a sacrifice to the idol of Zeus and, in doing so, set off a firestorm that changed the physical, political, and religious landscape of the Levant for the next hundred years.

In military terms, what Mattathias, his son Judah, and Judah's brothers accomplished was remarkable—if not miraculous, as it is represented in the Hanukkah story. The Maccabee ("hammerer") brothers and their small band of followers took on a vast, well-trained Seleucid army that was equipped with the best weapons and even battle elephants. Yet within three years (by 164 B.C.E.), the Jews had recaptured Jerusalem and reconsecrated the Temple. The next twenty-five years saw continued struggles between the Hasmoneans (named for an ancestor of Mattathias) and the Seleucids. Finally, in 141 B.C.E., Simon, the last of the Maccabee brothers, became high priest and political leader, the first of the Hasmonean priest-kings.

With the rise of the Hasmoneans, the Cohanim were now in charge. Not only did they sit on the throne, but they were also at the top of Jerusalem society. As Lee Levine observes in *Ancient Israel,* priests now played central roles in administering religious, political, diplomatic, and military affairs; for the time being, they had eclipsed the kings, sages, and prophets. Jerusalem and the Temple were again the focal point of global Jewish worship and culture. The Hasmonean period was also the last time that Judeans independently ruled any part of their ancient kingdom until David Ben Gurion declared the state of Israel two thousand years later.

As time passed, Rome became more unsettled by the power of the Hasmoneans. But as it happened, the Romans had no need for their unrivaled military machine to gain entrée into Israelite affairs: as was too often the case, trouble began within. The Romans came to Jerusalem by virtue of an invitation to settle a Judean family feud. The last vestiges of the Hasmonean dynasty, the brothers Hyrcanus II and Aristobulus II, each lobbied the Roman legate in Syria for recognition as the true ruler of Judea. When Rome threw its support behind Hyrcanus, Aristobulus's zealous faithful barricaded themselves inside the Temple for three months. In 63 B.C.E., the Roman general Pompey conquered Jerusalem and greatly curtailed the Jewish territory ruled by the high priests, effectively wiping out the Hasmonean Empire. For the larger priesthood, the arrival of the Romans was no less catastrophic. The Cohanim lost their political independence and most of their power; once again their purview was limited to the sacred.

The Romans not only reined in the Cohanim's influence but also installed pro-Roman priests who were not of priestly stock. As Shaye J. D. Cohen notes in *Ancient Israel*, opponents of these priests claimed that many were not even Jewish. This trend toward political loyalty as the primary criterion for priesthood (rather than heritage or professional qualifications) reached its peak under Herod, whose family had converted to Judaism three generations earlier and who ruled the Jews from 37 to 4 B.C.E. Herod purged both the Temple and the city of all remnants of the Hasmonean dynasty. He favored Jews from the Babylonian and Hellenistic Diasporas in both his priesthood and his court. Moreover, for a time he dissolved the Sanhedrin, the council of Jewish sages who constituted the highest judicial and legislative bodies in Judea during the Roman period.

Despite their obvious weakness and subservience in comparison with their counterparts of the Hasmonean era, Rome-friendly priests occasionally influenced religious events that rippled far beyond the confines of the Tabernacle. Their involvement in the trial of Jesus of Nazareth is the most infamous example. As recounted in the New Testament Gospels (Matthew 26:62–66, Mark 14:60–64, Luke 22:66–71), Jesus was brought before the Sanhedrin, questioned by the high priest, and charged with blasphemy. Jesus's novel interpretations of the Torah and his excoriation of Jewish "scribes and Pharisees" as hypocrites sentenced to hell cannot have won him many friends among the Judean religious establishment trying to hold on to power during an especially turbulent and fractious time. The Sanhedrin handed him over to Pontius Pilate, the Roman prefect, who carried out his crucifixion.

Momentous verdicts of the Sanhedrin notwithstanding, by the first century C.E. the priesthood's power continued to wane. Not only had it been diluted by Roman hegemony, but the practice of Judaism had also begun to move out of the Temple and into homes and synagogues, where prayer and Torah study were increasing. This focus on the individual rather than on the tribe, clan, or community necessarily came at the expense of the priest. Quite apart from the ascent of Christianity during this time, Judaism itself—already fractured into three major sects (Pharisees, Sadducees, and Essenes)—further splintered into a variety of revolutionary and apocalyptic groups. As Shaye Cohen observes in *Ancient Israel,* the growth of these movements signaled a fundamental breakdown of the social and religious order.

This breakdown was completed with the "Great Revolt" of the Jews against the Romans in 70 of the Common Era and the destruction of the Second Temple. The social and religious unrest of the post-Herodian period, coupled with Roman cruelty and capriciousness, was the main antecedent. The priesthood was hardly exempt from the unrest: there was now ongoing internecine competition—and sometimes physical violence—between upper and lower clergy. In the early part of the rebellion, a group of aristocratic priests, led by Eleazar, was at the leading edge of the fighting against the Romans.

In the spring of 70 C.E., Emperor Vespasian's son Titus besieged Jerusalem. The Romans' primary military and symbolic target was the Temple itself, where the priests continued daily sacrifices. On the tenth day of the Hebrew month of Av, Titus's forces destroyed the Temple.[2] Beyond a "Jewish tax," however, Titus did not impose any other punitive measures on the Judean populace. Jewish life was therefore allowed to persist, albeit in a different and reduced form.

For all practical purposes, the destruction of the Second Temple marked the end of the Jewish priesthood. Forces undermining the authority of the Cohanim, however, had long been in place. Wars, politics, and economic opportunity led many Second Temple–era Jews to move, taking their religion with them. In practice this meant the emergence of a less centralized system of worship, including prayer instead of animal sacrifice in a single Temple and personal responsibility to obey the commandments of the Torah instead of a central institutional intermediary guiding religious practice.

The absence of sacrifices, the end of a Jewish polity, and the rise of a more individual Judaism rendered the priests superfluous. The rabbi—part scholar, part theologian, part teacher, part social worker—became the new model for Jewish clergy as two millennia of Diaspora life unfolded.

The Priests Today

Although the Jews have had no high priest for almost two thousand years, there are still individuals who consider themselves priests today: a sizeable minority of Jewish males consider themselves to be Cohanim, directly descended from one of the many priests who served in the Temple in Jerusalem. Today, presumptive priestly descendants bless their fellow Jews at the Western Wall and in synagogues on the High Holidays, on the three pilgrim festivals, and, subject to local custom, on other occasions as well. The split-fingered gesture that accompanies their benediction, however, now blesses more than just the chosen people: *Star Trek* aficionados throughout the United Federation of Planets recognize the same hand symbol as the Vulcan greeting meaning "Live long and prosper!" The symbol was imposed on the first *Star Trek* series by Leonard Nimoy, the Jewish star who came to be inseparable from Captain Kirk's loyal partner, Spock. Self-identified Cohanim include a hand symbol as a crest on their gravestones, a visible sign of their priestly status (and a useful one for geneticists, as we shall see).

The presence of Cohanim today is a powerful testament to the enduring abilities of the Bible and oral tradition to shape the modern world. But how is it that nearly two thousand years after the Cohanim lost their jobs in Jerusalem, there are still priests around to perform ceremonial duties?

Let's consider what we know: Cohanim served in the Temple in Jerusalem more than two thousand years ago (and perhaps as far back as three thousand years); unlike Jewishness itself, which is traditionally passed down through the mother (matrilineal descent), priestly status was conferred only on males, passing down from father to son (patrilineal descent); and there are male Jews today who regard themselves as Cohanim, direct male descendants of those priests who served long ago. So what happened in the intervening centuries?

We can imagine two very different sorts of answers: adoption and genetic continuity. The Jews have long carried with them the story of the priests, and at some point in their history these stories could have motivated people to assume, or adopt, priestly status, regardless of their genetic ancestry. In other words, sometime after the dispersal in the first and second centuries C.E., a group of Jews (or maybe non-Jews, a possibility I consider in chapter 3) could have decided to adopt the title of priest or been awarded it. In time, this group could have come to be accepted as such. We know that in the Second Temple period the award of priestly status had everything to do with political expediency and relatively little to do with genealogy.

At the other extreme from this nongenetic adoption of status is the possibility of genetic continuity, which assumes that the oral tradition is largely correct: the Cohanim of today are indeed culturally and genetically continuous (on the male side) with the priests who survived the destruction of the Second Temple and subsequent uprisings. Such continuity is not unprecedented. In fact, the Levite genealogy presented in the book of Chronicles (1 Chronicles 6:1–15) traces a continuous priestly line of descent from Aaron to Jehozadak, whose son Joshua became the first high priest of the Second Temple era.[3] Nevertheless, the possibility that a similar continuous line might be traced from the time of the destruction of the Second Temple in 70 C.E. down through two thousand years of turbulent history seems unlikely on its face. And like most of the rest of human history, the continuity question has been beyond the reach of science.

That is, until now.

The Genetic Lineage

As Neil Bradman and I discussed in the late 1990s, assessing the lineage of the Jewish priesthood is well suited to genetic analysis. In general, genetic history has been successful only in contexts where nongenetic sources of information—for example, texts, oral history, or archaeology—specify mutually exclusive possibilities. In this case, both textual sources and oral traditions of the individuals who consider themselves priests allow us to specify the alternative scenarios: adoption versus inheritance of priestly status. Working without such external information, it would be difficult, if not impossible, to tell a coherent story about the history of a population based solely on genetic data. We might be able to say that the population has or has not been the same size for a long period of time or that it has interbred with another population. But even these general inferences would need to be qualified.

On the other hand, disciplines outside genetics often give us very precise alternatives to consider. We have historical accounts of both continuity and adoption of priestly status. Might genetics support either possibility? To find out means to study the Y chromosomes of putative Cohanim, which is what Neil, I, and our colleagues did.

Before we delve into the details of the genetics, we need to consider the Y chromosome, the linchpin of this project. Among some geneticists the Y chromosome gets a bad rap: it's small, carries few genes, possesses little variation, and, if some predictions are correct, appears to be degenerating into a genetic wasteland (because of a lack of genetic interaction, or recombination, with other genetic material, it has been considered prone to deterioration over evolutionary time). All of that may be true (we're not sure about the last part), but for genetic historians the Y is indispensable.

The Y chromosome is inherited through the paternal line. Human males have one whereas human females have none. A father passes his Y on to his sons, never to his daughters. Moreover, because the Y does not undergo the evolutionary shuffling process that the paired chromosomes (1 through 22 and, in females, X) do, sons inherit a Y from their fathers that is essentially unchanged.

Of course, small changes in chromosomal DNA do occur. And it's a good thing for geneticists, too, because it is the interpretation of those changes as they are passed down from generation to generation that can help to illuminate human history. Polymorphisms—which can result from the small changes or mutations that occur from generation to generation—are present in part because so little of the Y consists of genes that actually do anything. Genes, generally speaking, are those segments of genetic material (DNA) that encode proteins that are responsible for much of an organism's physical and physiological characteristics. Most of the DNA on the Y (perhaps 99 percent) does not code for protein and we don't know what, if anything, it does. Such DNA is sometimes dismissively called "junk DNA," though increasingly genomicists recognize that functional parts of the genome are not restricted entirely to sequences that directly encode protein. It is nonetheless true that the sequence that does not code for protein is generally more free to evolve—that is, to acquire changes, or mutations, that lead to polymorphisms during the imperfect DNA replication process that are then passed on to succeeding generations.

From the perspective of genetic history this characteristic means Y chromosomes carry a lot of information about who is related to whom. As a Y passes through multiple generations, random changes occur in its junk DNA. Our Y chromosomes therefore retain a record of their passage through time: they can reveal the paternal genealogy of their owners and the genetic relationships among groups of individuals. A series of polymorphisms along a Y (or any other) chromosome may be looked at in concert and thereby provide more information about that chromosome's history; such a group of polymorphisms is referred to as a *haplotype*. By making assumptions from haplotype data about rates at which different types of mutations occur, it is possible to estimate a date for the most recent common ancestor of any two or more Y chromosomes.

The Y chromosome therefore is particularly suitable for investigation of the genetic history of the Cohanim. Do those people who claim to be descended from Old Testament priests have a legitimate case for genetic descent as opposed to simply cultural identification? If so, it would have to be recorded in some way in their Y chromosomes. If priestly status were adopted without regard to genealogy, there can be no such thing as a hereditary priest. If, on the other hand, the Cohanim have followed the oral tradition of paternal inheritance, then there must be signatures recoverable in their DNA.

Let us examine the easy case first. Assume that no historical basis exists for the story. Assume that either in the recent past or over many generations, some fraction of the Jewish population randomly adopted priestly status. Assume further that the adoption was, in a genetic sense, completely independent of the priests who once served in the Jerusalem Temple. And assume, finally, that a large number of such random adoptions of priestly status occurred. In this scenario, what would the Y chromosomes of the priests look like in comparison with those of the general Jewish population? In short, we would expect them to look the same on average. But what would we expect if the oral tradition surrounding the priesthood is largely correct?

Before we could answer that question, we needed to have an idea, however crude, as to what proportion of Jews might be considered Cohanim. My solution was to keep reading books in the hope that someone had mentioned the issue somewhere. Neil Bradman came up with a better idea: go into Jewish cemeteries and start counting all the headstones with the priestly hand symbol on them. Visit enough cemeteries, reasoned Neil, and you'd get a good idea of the proportion in the general population. In this way, Neil made a casual estimate based on Cohanim symbols found in cemeteries in the United Kingdom and elsewhere. I myself spent a brisk morning wandering around the haunting and beautiful Jewish cemetery in Prague searching for the Cohen hand symbol and feeling a surprising sense of connection to this community of which I knew so little. Could some of the inhabitants of these Czech graves have been descended from ancient Hebraic priests?

By tallying hand symbols, Neil and his coauthors of the original *Nature* study arrived at an estimate: 4 or 5 percent of male world Jewry are "priests." Using that number, we estimated that as many as 500,000 Cohanim exist among the Sephardi and Ashkenazi Jews in the world today.[4] According to tradition, therefore, the 500,000 Y chromosomes carried by the Cohanim of both contemporary communities are derived from a single ancestral chromosome some time in the past three thousand years or so.

If the oral, hereditary tradition were true, even partially, then the Y chromosomes of the Jewish priests would not only be different from those of the general Jewish population but also would be different in a very particular way: they would be much less heterogeneous. In the general Jewish population, we would expect many different types of Y chromosomes: those of Jews and converts to Judaism from all over, representing all different occupations and genetic origins. However, in the priests we would expect few types of Y: they would constitute a smaller, more homogeneous population representing just one occupation with a strong and perhaps unique hereditary component (at least with respect to ancient times). Those few types of Y would furthermore be very similar to one another, because they would all have derived from a single common type in the past.

To compare the Y chromosomes of the priests and the general Jewish population, our team needed several other types of information. The samples in this case were mostly already available. Our collaborator Karl Skorecki had collected samples in Canada, while Neil's son Robert had taken on the hard job of swabbing the DNA from the mouths of sunbathers on the beaches of Tel Aviv. The rest came from the United Kingdom. Care was taken to ensure that the individuals were not related, at least to the grandfather level on the father's side. This was important because one large extended family, if all sampled, say, within the Cohanim, would have generated the impression that all Cohan Y chromosomes were similar.

The priestly or lay designation was based entirely on self-identification, independent of surname. Surnames are obviously correlated with priestly status. Individuals named Cohen, Coen, Kagen, and Kohn are often considered Cohanim; indeed, these individuals often identify themselves as such. But surnames are not a sure guide to priestly status. Cohens are often Cohanim, but not always. And individuals with other surnames not obviously connected to Cohen may be Cohanim. By custom the designation is based on self-identification. In testing the oral tradition, we wanted to look not at surnames but at whether people considered themselves priests, just as we looked for the hand symbol on gravestones rather than surnames. Thus, volunteers were asked whether they were Cohanim, Levites, or Israelites.

To increase the power of the analysis, volunteers were also asked if they were Ashkenazi Jews, descendants of northern European communities, or non-Ashkenazi Jews, meaning in this case descendants of North African or other Near Eastern communities such as Morocco, Iraq, and Iran. Dividing the volunteers into Ashkenazi and non-Ashkenazi categories allowed us to compare priests and Israelites independently in each community. Because these communities (defined by the practice of different religious customs) have been separated demographically for at least five hundred years and probably much longer, repeating the analysis within each community gave us considerably more power to distinguish between "priestly" and "lay" chromosomes. A pattern seen in both could not be very recent.

Once the DNA had been collected and prepared for analysis, we needed a way to distinguish different types of Y chromosomes. For this study, we used a set of six unique event polymorphisms, referred to here as "unique mutations" for convenience. Unique mutations are usually just single-base substitutions in the DNA four-letter alphabet code (A, T, C, and G—known as "bases"). Thus, a unique mutation might be a T replacing a G. These sorts of mutations occur very rarely. Therefore, when we see multiple Y chromosomes carrying the same mutation, we can be just about certain they have a common ancestor. The rare violations of this rule become readily apparent by analyzing combinations of "unique" mutations. These unique mutations, however, usually identify relatively large groups of related chromosomes. Normally we also require a method to identify relationships among chromosomes *within* these large groupings. For this purpose, we use a special class of polymorphism I've already mentioned called microsatellites, the subject of my work on human evolution with Luca Cavalli-Sforza and Marc Feldman.

Microsatellites are not small communications devices. Rather they are a type of repetitive DNA whose repetitions frequently vary from person to person. When short motifs of DNA bases, such as CA or CAT, are repeated in tandem, the machinery that copies DNA from generation to generation often trips up and gets the number of repetitions wrong. Ten consecutive repetitions of CA might become eleven repetitions. For most normal regions of DNA, without such repetitions, the machinery that copies DNA is extremely efficient and accurate; its error rate is less than one in a hundred million. But these repetitive stretches, or microsatellites, trick and confuse the machinery to such an extent that errors can occur up to one out of every thousand or so tries. In essence, the copying machinery ends up miscounting the number of repetitions, mistakenly increasing or decreasing the total number. What this means is that even among groups of Y chromosomes that are all closely related to one another (for example, two groups that both carry a particular mutation), we can still make fine distinctions among them in their exact relationships —microsatellite mutations occur often enough to allow us to make those distinctions.[5]

The first and simplest question we could ask was with the unique mutations. How many different groups of chromosomes could we find in the Cohanim and other Israelites (Jews who are neither Cohanim nor Levites)? In addition to this question about diversity, we could also assess the degree of genetic divergence between the Cohanim and the Israelites.

In the first study, led by Neil, Karl Skorecki, and Mike Hammer, chromosomes were classified into three groups defined by the unique mutations and one microsatellite. These three groups of chromosomes showed very different frequencies between the Cohanim and the Israelites. For the Cohanim, more than 90 percent were in what was called group 1, whereas for the Israelites only 62 percent were. Thus, this first study suggested that there was something unusual going on. But what exactly?

Once a battery of the more quickly evolving microsatellites was used to identify finer relationships among the chromosomes, the differences between the Cohanim and Israelites became even more striking. Using a combination of the two types of genetic markers (unique mutations and microsatellites), we identified 109 different types of Y chromosomes in a total sample size of 306 men. Now, 109 different types is a lot of variation, especially in a sample of little more than 300. With so many chromosome types represented, it would be very easy for the data to be essentially noise: a few of this type, a few of that type. In that case, we would have to conclude that priesthood was probably a function of adoption, not genealogy or oral tradition.

But here's where hard work paid off. Despite the high level of variation, we could see a clear difference between Cohen and Israelite chromosomes. The most common chromosomes observed in the Israelites (that is, non-Cohen and non-Levite Jews) were found in only 12 percent of the Israelite individuals sampled. By contrast, more than half of the Cohen Y chromosomes were identical at all the sites considered—that is, the majority of the self-identified Cohanim had the same type of Y chromosome. Even more remarkable, this same type of Y was found at high frequencies in both Ashkenazi (45 percent) and Sephardi (56 percent) Cohanim.

We named the chromosomal type found in more than half the Cohanim the Cohen Modal Haplotype. *Modal* means "most frequent," and *haplotype* indicates a particular constellation of mutations (or, more accurately, variant forms) at all the sites, or markers, being considered.

As it turned out, there is not only a single type common among the Cohanim but, critically, also many chromosome types similar to the most common one. If we consider all the chromosome types that are clearly related to the Cohen Modal Haplotype through a small number of mutations (that is, changes in size at the fast-evolving microsatellites), we can identify a set of related chromosomes that we call the modal cluster. Among the Cohanim, the modal cluster accounts for 64 percent of the observed chromosomes. Among the

Israelites, this cluster is again the most frequent, but it accounts for only 14 percent of the chromosomes observed. This, too, makes clear how similar the Cohen chromosomes are to one another in comparison with the Israelite chromosomes.

Just as with the Cohen Modal Haplotype frequencies, the modal clusters found in the Cohanim of the Ashkenazi and Sephardi communities are remarkably similar: 69 percent of Ashkenazi and 61 percent of Sephardi Cohanim share haplotypes found within the modal cluster. These similarities are particularly striking given the length of time the Ashkenazi and Sephardi communities have been separated. The similarity of the Cohen chromosomes in both communities therefore appears to reflect an old lineage that predates the separation of these communities.

So with this simple comparison we had already learned something about Jewish priests that no one knew before. Even allowing for periods when conferral of priesthood was based on political expediency, the evidence for widespread arbitrary adoption, in which many Jews simply took on priestly status, does not hold up under genetic scrutiny. Given the relative homogeneity of Cohen Y chromosomes in comparison with those of the Israelites, we can conclude definitively that adoption of status has not occurred on a very large scale over a long period of time. In fact, even a little bit of arbitrary adoption each generation would erode the signal because it would propagate over time; similarly, even infidelity on the parts of the wives of priests would erode the signal. Calculations have been made, in fact, suggesting that what is called the nonpaternity rates for children of priestly couples must be considerably less than the 5 percent or more often estimated in modern society.

Does all this mean that we have validated the oral tradition? Not quite. Unfortunately, the real world is more complicated. From our two original possibilities of complete continuity from a single, ancient founding priest or large-scale arbitrary adoption, we have rejected the latter. But that does not prove complete continuity. Another possibility is that Cohen and Israelite Y chromosomes are different because a small number of individuals recently—say, in the last few hundred years—adopted priestly status (presumably before the separation of the Sephardi and Ashkenazi communities, unless we are prepared to entertain the possibility that both groups independently established the same dominant chromosome type). In a scenario of a more recent origin in a small number of related individuals, current priests would trace back genetically to this group of men in a line no less straight than the one traced from Aaron to Jehozadak by the Old Testament Chronicler (1 Chronicles 6:1–15).

One possibility is that this adoption occurred more or less concurrently with the adoption of surnames in Jewish populations beginning in the late Middle Ages. But even if we proved continuity over a three-hundred-year period, that still would not settle the question—surnames went back at least that far, perhaps farther. Thus, genetic continuity back to an eighteenth-century eastern European shtetl is quite different from genetic continuity back to priests sacrificing animals in the Jerusalem Temple during the time of Christ. Surname continuity might be consistent with priestly adoption, but it could not definitively prove or disprove it.

Although we could certainly still publish a simple correlation between priestly status and Y chromosomes because of the strong impression of a biblical connection, without dating the lineage we can't really get beyond the very recent past. And even having found a characteristic chromosomal type, which we called the Cohen Modal Haplotype, we would not have added much to the work Karl Skorecki, Mike Hammer, and their collaborators had already published in *Nature*. We would still be left with a set of similar chromosomes common to presumptive Cohanim, but there would be nothing to prove that those weren't simply the by-product of, for example, Kohn-to-Kohn or Kahne-to-Kahne transmission over the last few generations. We therefore decided to use genetic tools to take a different approach: to see if we could put a date on the ancestral Cohen chromosome. If it turned out to be only a few hundred years old, the adoption case would have to be seriously reconsidered. But if it turned out to be much older, the evidence for the oral tradition would become that much stronger.

Dating "Aaron's Lineage"

We had already established that the Y chromosomes carried by Cohanim are sharply different from those found in the general Jewish population. The majority of priestly chromosomes are closely related to one another, showing clear evidence of common descent from a single ancestral Y chromosome at some point in the past, which we would like to date. The type of chromosome responsible for this particular lineage is visible as the modal cluster found in more than 60 percent of today's Cohanim.

It is especially significant, however, that we had a *cluster* of closely related chromosomes, and not just a single type. If the ancestor of the chromosomes observed in the Cohanim were very recent, we would not see a cluster of related types—there would not have been enough time for the single ancestral Y to give rise to even a modest number of variants. Given the genetic resolution we were using, we would see exactly one type.

As a simple illustration, imagine that we could look at the Y chromosomes of the grandsons of Meyer Rothschild, founder of the European banking dynasty. Using the set of markers we applied in the Cohanim study and assuming that the Rothschild wives did nothing untoward, the Rothschild descendants should all have the same type of Y chromosome—the one carried by Meyer Rothschild himself. Two generations is not enough time for mutations to occur at the few places on the Y chromosome that we inspected, even for the rapidly changing microsatellites. Since we observed multiple closely related chromosome types among the Cohanim, however, we can conclude that the Cohanim line was established more than two generations ago. If we could find out how much more, we might be able to determine if there is genetic continuity back to the age of surnames or much more ancient times—closer, say, to the age of the Temple.

To arrive at an estimate of how long ago the Meyer Rothschild of the priestly line might have lived, we needed a genetic model. I spent a long, frenzied weekend in my small, terraced house in north Oxford thinking about what that model might look like. Whatever it was, it had to take into account two things: the genealogy and the mutation process. I needed to use assumptions about those things to estimate when the man lived who carried the ancestral chromosome and founded the Cohen line. To see how I estimated how long ago the founder of the Cohen line lived, it is necessary to understand what is represented in the genealogy (figure 2). Imagine again a male ancestor and a number of his descendants (for convenience, each one has two of his own). Tracing the process forward, we have the ancestral chromosome and all of its descendant chromosomes (labeled "ancestor" and "descendant" in the figure). By generation 4, there have been three opportunities for mutation in each of the indicated lineages leading to each of the chromosomes present in the descendants. This is not much time, but there may have been, say, a single mutation at a rapidly evolving microsatellite. As more time goes by, more and more mutated chromosomes appear. This is how I related mutational differences to time. I should note here that this figure simplifies the real picture by using an example where the relationship among the chromosomes present today (generation 4) is the same as the genealogy for the individuals, which is not generally the case.

With most genetic data, including microsatellites, it is not simply a matter of counting up observed mutations the way one would gravestones. Unfortunately microsatellite mutations can disappear by overwriting

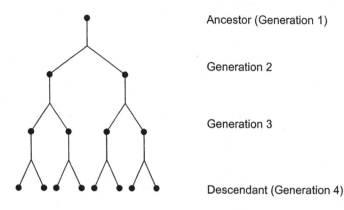

Ancestor (Generation 1)

Generation 2

Generation 3

Descendant (Generation 4)

Figure 2. A genealogy showing the inheritance of a unisexual genetic system, for example, the Y chromosome

themselves like a palimpsest. Because they mutate by changes in length (and over a finite range), a step up may be followed by a step down, and it will appear as if no mutations have occurred at all. Thus, to arrive at an appropriate time estimate, we had to devise a mathematical trick, or correction, that could count up these "missing" mutations. This correction can be derived under more or less complicated scenarios, though we chose one of the most simple approaches.

To arrive at our estimate of the age of the Cohen Y chromosome lineage, we needed one more step. Neil and I knew how to estimate the time separating two chromosomes. But what we were really interested in was the time from the bottom of the inverted Cohanim family tree back to the top—the years separating the chromosomes being sunned today on the beaches of Tel Aviv and those of the founder of the line, who directed processions in the Jerusalem Temple or slaughtered animals before the Tent of Meeting in the Sinai desert. The approach I developed was the simplest one possible: it amounted to making an educated guess as to what the ancestral chromosome was and then calculating the distance from the current chromosomes to this imagined ancestral one. In the case of the Cohanim, the guess was an easy one: clearly, the ancestral chromosome was the Cohen Modal Haplotype, because it is by far the most common. Thus, Neil and I needed only to count the number of differences separating current chromosomes from the Cohen Modal Haplotype, to average this figure over all the chromosomes, and then correct for all the hidden changes.

That was the theory, at least. I wrote a computer simulation to see if I was right. Being a lousy mathematician, I always had to "run out" the process on a computer and check whether my equations (guided usually by a thought experiment instead of mathematical rigor) gave the correct answer. The simulations showed that the method worked fine as long as you knew the ancestral chromosome type.

So now it was time to apply it to the data. From my office in the Zoology Department at Oxford University I called Neil, who was at his house in Herzliya Pituach, a coastal town north of Tel Aviv. I told him what I wanted to do, and he put the phone down and came back minutes later with all of his data sheets in front of him. Thus, with twenty-two hundred miles between us, we began to analyze the microsatellite data that Mark Thomas had given Neil. One by one, Neil read the microsatellite scores for each chromosome carried by the Cohanim. As he spoke, I entered each number into the equation and tallied up the totals. Slowly the data began to take shape, and as they did, the presumptive date for the Cohen Modal Haplotype began to recede farther and farther down the corridors of time. "Do you know where we are going?" I said, feeling like a window had opened from the modern world into something ancient, powerful, and hidden.

When the last figure had been entered, we were stunned into silence. "We are in the First Temple," I said eventually. We were both quiet again for a time. When we accounted for the "hidden" mutations in the microsatellites, the figure we got is about three thousand years before the present, or right about the time that Solomon is thought to have been building the Temple in which the priests would serve.

Permit me here, after what was for me the first—and still one of the few—real thrills of discovery that punctuate the tedium and detail of science, the necessary reality check. Our results appeared to be a striking confirmation of the oral tradition. It even led to repeated claims in the press that my colleagues and I "found Aaron's Y chromosome." But although three thousand years is our best guess, the range of possible dates was and is very broad. Given our uncertainty about the ways mutations happen and how fast, we may be off by several hundred years or more in either direction.

Looked at this way, the fact that our date corresponds to the date of the building of Solomon's Temple may very well be coincidental. But if the range of possible dates is so broad, why calculate a date at all? While the origin of the line cannot be assigned with precision to the time of Aaron, Solomon, or any other named individual, we can rule out a very recent date. That is, we know now that the origin of the line is older than the development of surnames in Jewish populations.

The Aftermath

I keep a folder of the various press clippings that mention my work, mostly because my mom and dad often send them to me but also because they are a powerful reminder of the ways in which even modest and speculative research findings get exaggerated and reported as fact. For me, the Cohanim study and the "Aaron's Y chromosome" headlines it generated represent one of the more egregious examples of this phenomenon. "Jewish line traced back to Moses," wrote a reporter in the *Independent* (9 July 1998). The *Jerusalem Report* (10 May 1999) declared a little more reasonably that we had found "scientific confirmation of an oral tradition passed down through 3,000 years." I understand the need to sell newspapers, but these unequivocal proclamations irritate me. In my view, it is striking enough that the Y chromosome could provide any historical information at all. I thought the story was that genetic history was not total bunk and indeed could add to conventional history, archaeology, and anthropology. But the press has a weakness for grandeur.

What of the subjects themselves? What did they make of the results? Most were as flabbergasted (or gobsmacked, to use the infinitely more expressive British expression) as we were. Most, however, wondered why we'd bothered: "Well, naturally we have priestly chromosomes—we told you we were priests, didn't we?" I have no doubt many think they are indeed carrying Aaron's Y.

Could they be? There's no way to know, but while we have certainly not shown that to be true, neither can I or anyone else say it is impossible. According to the book of Numbers (20:22, 33:38), Aaron died at Mount Hor in the fortieth year of the Exodus. The Exodus cannot be dated with any precision, but according to John Bright's *History of Israel,* the totality of the textual evidence suggests that if the Exodus occurred at all, it was likely some time in the thirteenth century B.C.E. This date, if indeed it coincided with Aaron's life, would be well within the interval we predict for the origin of the Cohanim.

Genetic history is still history—in the absence of a time machine, we will probably never know when Aaron or his descendants first donned vestments, made sacrifices, and consulted the Urim and Thummim (or whether they ever did). What we can say is that our best guess is that the priestly line was founded before the major dispersals of the Jewish populations—that is, certainly before the time of the Romans and perhaps before the Babylonian conquest in the sixth century B.C.E. Whatever their true origins, whomever they may trace to, the Cohanim of today are descended from an ancient lineage.

CHAPTER 2

Lost Tribe No More? The Black Jews of South Africa

These are the children of the province,
who went up out of the captivity,
of those that had been carried away,
whom Nebuchadnezzar the king of
Babylon had carried away, and came
again to Jerusalem and to Judah,
every one unto his city. . . . The
children of Sena'ah, three thousand
nine hundred and thirty.

Nehemiah 7:6, 38

Years ago, there was an ad campaign that featured a Japanese boy holding a piece of bread. "You don't have to be Jewish to love Levy's real rye," it declared. The campaign was hugely successful—the catchphrase "You don't have to be Jewish to" has since been attached to countless other verb phrases: "eat kosher," "believe in Jesus," "love JDate," even "be a neocon."

I would add one other: "become one of the world's foremost experts on the anthropology of the Jews." I'm talking about Tudor Parfitt, professor of modern Jewish studies at the University of London's School of Oriental and African Studies, just down the road from where we were toiling at UCL. Not only does Tudor, born and raised in the Church of England, speak fluent Hebrew, he's forgotten more about Jewish and ancient Israelite history than I'll ever know. He was an acquaintance of Neil's, and Neil thought he might be just the person to help us determine whether genetics might verify whether a certain tribe in Africa claiming to be a Lost Tribe of Israel is correct. Tudor Parfitt indeed proved indispensable to this project.

Perhaps no Old Testament legend has been retold more often than that of the Lost Tribes. Jacob, the last of the patriarchs, had twelve sons. These were the presumptive ancestors of the tribes of Israel and the ones for whom each tribe was named. Following the death of King Solomon in the tenth century B.C.E. and the conquest of the Northern Kingdom of Israel in the eighth century B.C.E. (by the Assyrians), the ten northern tribes were ostensibly "lost." Since then, at one time or another innumerable peoples from all over the world—Japanese, Native Americans, Indians, Persians, Ethiopians, North African Berbers, Eskimos, and Tauregs of the Sahara, among others—have claimed descent from one or more of the Lost Tribes.

Genetic history is unlikely ever to tell us if we've found actual descendants of a Lost Tribe. What we can do with genetics, however, is evaluate whether groups showing Jewish cultural characteristics or claiming Jewish ancestry show genetic affiliation with other Jewish groups, which is of course a more narrow and specific question. That type of evidence will never prove one population's claim to Jewish ancestry; nor will it ever rule out another's.[1] But it can give us clues about a people's origins and migrations. And genetic history does not take place in a vacuum—when scientists make statements about how closely one group is related to another or inferences about a people's genetic ancestry, there are real-world reverberations.

In this chapter I first review the origin of the Lost Tribes story. I then explore what genetics tells us about the Lemba, an enigmatic group from southern Africa claiming ancient Judaic connections. Finally, I comment on the consequences of our group's study for the Lemba's self-identity and broader questions of genetic history and group consciousness.

In recounting the story of the Lost Tribes, I do not mean to lend credence to one or another population's claim to being a lost tribe. Rather, I hope to convey how the Lost Tribes myth came to be so pervasive and so resonant among contemporary populations and how it continues to resonate in popular perceptions of my work. Why do so many lay claim to this story?

Recall that every tribe was given its own separate territory, except the tribe of Levi, which was designated for priestly duties. There were other lineup changes among Jacob's sons, too. For example, Reuben lost his rights as firstborn son, perhaps understandably, by committing adultery with his father's concubine (Genesis 35:22, 49:3–4). Joseph was replaced by his sons Ephraim and Manasseh (Genesis 48:5–6). Meanwhile, Simeon, Levi's co-conspirator in the slaughter of Shechem and his Canaanite village (in revenge for the rape of their sister), had his tribe absorbed into that of Judah long before any of the other nine northern tribes were lost.

In any case, somewhere around 1050 B.C.E., the twelve tribes coalesced to form the Israelite Kingdom ("United Monarchy") under Saul. According to the traditional view, his successor and son-in-law, David, further consolidated the tribes within the monarchy, slew various enemies, and grew the kingdom into an empire extending from modern Lebanon/Syria in the north to the Sinai desert in the south, and from the Mediterranean coast in the west to the edge of the Arabian Desert in the east.[2]

David's son Solomon oversaw the construction of the First Temple and presided over an unprecedented economic boom characterized by peace and prosperity. After Solomon's death, however, his heirs could not hold the monarchy together. In 920 B.C.E., the Israelite Kingdom was split into two: Israel in the north (containing the nine northern tribes) and Judah in the south (formed from the territories of Judah, Simeon, and Benjamin).[3] In 722 B.C.E., the Assyrians finished conquering the so-called Northern Kingdom of Israel (the process had begun a decade earlier), laid waste to the capital of Samaria, and sent countless Israelites into exile. The elite ruling class of the Northern Kingdom was deported to different parts of the Assyrian Empire and summarily replaced by a new set of exiles imported by the Assyrians. The Assyrians may have been the first to systematize ethnic cleansing, and they went about it with staggering efficiency. When cities and even kingdoms were cleansed of their leading citizens by the Assyrians, they simply disappeared from history. Had the Assyrians done the same to Judah at this time, it is doubtful that we would know much of anything about the ancient Hebrews.

The early United Monarchy has left us little in the way of direct and unequivocal archaeological evidence; by contrast, the facts of the Assyrian conquest of the Northern Israelites are disputed by no serious authorities. By this point in the history of the Israelites we are on solid archaeological ground and we need not depend on the Bible, as we must for so many of the details of the earlier period. Both the destruction of Israel and the subsequent population transfer are well documented by Assyrian archaeological and epigraphic evidence. For example, battles between the Assyrians and the Hebrews are commemorated in remarkable detail in the British Museum, just down the road from where I worked in London. I am always struck by one display in particular—a set of Assyrian bas-reliefs depicting columns of Assyrian soldiers leading captives away from the defeated city of Lachish, one of Judea's largest enclaves, attacked in a later Assyrian campaign. The Assyrian soldiers all wear exactly the same countenance, leaving a chilling impression of an army of immutable resolve, no doubt the intention of the imperial artists.

With the fall of Samaria to the Assyrians, the northern tribes are said to have been "lost"—the Old Testament has little else concrete to say about the fate of the exiled Israelites. Most scholars agree that, in all likelihood, the exiles were simply assimilated by their Diasporan hosts, principally the Medes, an Iranian people who inhabited what is today western and northwestern Iran. In addition, some Israelites undoubtedly fled south to Judah, where they were sure to find a familiar language, religion, and way of life.

As Hillel Halkin observes in *Across the Sabbath River,* the Bible's silence on the precise fate of the exiled Israelites does not imply a lack of interest in their fate. Rather, what happens after the book of Kings is that the Lost Tribes become the subject of prophecy: for example, in Jeremiah we are told that when God ultimately reconciles with His chosen people, "He that scattered Israel will gather him, and keep him, as a shepherd doth his flock" (31:10).

Over time the legend of the Lost Tribes has only mushroomed. The first-century Jewish historian Josephus was convinced that the their descendants existed in great number "beyond the Euphrates." Meanwhile, countless charismatic figures have appeared over the last two thousand years, each claiming descent from one or the other tribe. Holy Roman Emperor Charlemagne (ca. 742–818 C.E.), for example, the first ruler of a great European empire since the fall of the Roman Empire centuries earlier, identified himself with King David. Not long after Charlemagne, a traveling salesman named Eldad HaDani appeared, claiming to be a descendant of the tribe of Dan. In his wanderings through the Middle East and Europe, he brought with him extravagant and fanciful accounts of the fate of the Lost Tribes as well as an ostensible body of rabbinic law (halacha) that was quite at odds with what was written in the Talmud. Eldad was regarded by many as a charlatan with messianic delusions; a number of later Talmudic authorities, including Rashi and Maimonides, however, considered his accounts authoritative. Whatever his veracity, Eldad and his claims of ancient Israelite descent made the Christian establishment nervous enough to construct the legend of Prester John, a mythical Christian king said to rule over a vast empire stretching from Africa to Asia whose subjects included the Lost Tribes.

Whole populations, too, have claimed lineage extending back to the Lost Tribes. Both the British and the Dutch have identified with them, and even members of the Japanese royal family have conjured some relationship. Native Americans, Pashtoons of Afghanistan, Tartars in Russia, and Maori in New Zealand—at one time or another, each has sworn kinship with the Lost Tribes.

Perhaps because it touches something deep within the human psyche —here I'm thinking of the powerful literary themes of home, the return of the prodigal child, grace, forgiveness, and redemption— the myth of the Lost Tribes has persisted. So seductive is the narrative that even some otherwise objective and deeply skeptical anthropologists confess a weakness for it. Tudor Parfitt, for example, admits that, when it comes to the Lost Tribes, he still regards himself as a "potential sucker."

All of which is to reiterate that, for me, genetic history is not about identifying a Lost Tribe but rather about the broader recurring themes of Jewish history contained within the myth of the Lost Tribes: exile, loss, wanderings, Diaspora, and reclamation of one's heritage. As much as any other piece of Jewish lore, the Lost Tribes story is about the continuing saga of strangers in strange lands.

Which brings us to the Lemba, a relatively small group of Bantu-speaking Africans living in South Africa and Zimbabwe who claim to have fled Judea three thousand years ago. Who are these people and how did they come to view themselves as descended from ancient Jews? To set the stage for the Lemba and their claims of Jewishness, let us first have a quick look at Bantu history.

Africa is to the study of human genetic diversity what the rain forest is to ecology. It is by far the most diverse and varied place on earth and also among the least well studied. Its many stories of human innovation and migration, of settlement and urbanization, remain all but untold. One dramatic episode in African prehistory that has been recovered to Westerners, thanks largely to the work of nineteenth-century linguists, is the stunning march of Bantu culture, language, and people across much of the African continent over the past four to five thousand years. The Bantu legacy extends from modern Nigeria and Cameroon in West Africa to the nation of South Africa, encompassing more than 200 million speakers of a Bantu or related language, amounting to nearly one in every three Africans.[4]

The linguistic and cultural heritage of this remarkable dispersal traces to a small expanse of savannah in what is now the border region of the modern states of Cameroon and Nigeria. There, between 3,000 and 2,000 B.C.E., a group of people speaking an early Bantu language developed distinctive and, for the time, progressive agricultural technologies that included cattle herding and farming of winter-rain crops such as African yams. These developments allowed the Bantu speakers to expand eastward, either directly through the rain forest or perhaps around its northern boundary. Somewhere in central Africa they split, with one group pushing farther south and another eastward. The eastern Bantu continued on to the Great Lakes region of Africa, an area that includes the whole of modern Burundi, Rwanda, and Uganda. There they found a thriving center of agricultural development. Around the time of the Second Temple, the Bantu speakers learned from the east African farmers how to work metal. The combination of winter-rain crop farming and metal work gave the Bantu speakers an overwhelming economic advantage in the tropical climates south of the Great Lakes. Consequently, the eastern Bantu speakers were able to sweep through Africa in one of the most rapid expansions in human history. By the time the British expanded their presence in South Africa in the eighteenth century, it was Bantu-speaking Zulu warriors who checked their advance.

Part of the cultural legacy of the Bantu speakers is the Lemba tribe of southern Africa. The approximately seventy thousand Lemba are currently spread through various parts of southern Africa, especially South Africa and Zimbabwe. By their own account, they arrived in southern Africa from Judea, "white men who came from Sena in the north by boat," half of them being lost at sea along the way. The Lemba regard the reference in the book of Nehemiah (see epigraph, above) to the children of Sena'ah's returning from the Babylonian exile as a direct nod to their ancestors. Lemba oral tradition is not very clear about Sena's location: sometimes it is in Judea, sometimes in Yemen or Ethiopia. The discrepancy may be explained by there being more than one Sena in their history: they speak of having rebuilt Sena after reaching Africa. "We are from Sena," Matshaya Mathiva, a Lemba scholar from South Africa, told Tudor Parfitt in the 1980s, "but we do not know where it is. We have lost the print of our foot on the earth."

Wherever that footprint might ultimately be found, one feature is clear. The Lemba consider themselves to be of Jewish descent, even though many today belong to Christian churches and others practice a brand of Islam. As Parfitt recounts in *Journey to the Vanished City,* the nearly universal refrain among the Lemba is some variation of "We are Jews, we came from Sena and crossed Pusela." (Pusela, like Sena, is a mysterious word; it is thought by many to mean the sea.)

Is there linguistic evidence the Lemba might be Jews? They speak a Bantu language but are also said to have a secret language called Hiberu. Might it be related to Hebrew? Like so much else of Lemba culture, the name of this secret language is a tantalizing, opaque, and somewhat misleading detail. As it happens, *Hiberu* is probably an ancient Egyptian word meaning "stranger" or "outsider"—its etymological connection to the word *Hebrew* is speculative at best. Hiberu the language is an antiquated form of Karanga, a southeastern Bantu tongue. And yet, it does contain some Semitic-sounding words, particularly tribal names.

However tenuous the linguistic link, there do appear to be circumstantial connections between the Lemba and Judaism:

- The Lemba are monotheistic—they do not practice idol worship. They claim that their God, Mwali, is the same as the God of the Israelites. Other African tribes also worship Mwali; the Lemba claim that they introduced the practice.
- Like Jews throughout much of history, the Lemba have resisted assimilation. They are endogamous—they do not take non-Lemba *wasenzhi* ("gentile") spouses. They keep to themselves and are secretive: Matshaya Mathiva compares them to the Freemasons.
- The Lemba practice circumcision. Indeed, a number of historical accounts suggest that the Lemba were the ones who introduced circumcision to sub-Saharan Africa.
- The Lemba are buried horizontally while most other African tribes bury their dead in a sitting position.
- The Lemba continue to practice animal sacrifice, complete with special knives and accompanied by prayers of thanks.
- Just as the Jewish calendar commemorates the new moon (*Rosh Chodesh*—"head of the month" in Hebrew), the Lemba make a point of knowing the precise details of the lunar cycle.
- The Lemba observe strict dietary laws somewhat akin to Jewish kashrut. Most notably, they do not mix milk and meat, nor do they eat pork.

Of course, none of these can be considered "diagnostic" of Judaism. Taken in the aggregate, however, they at least suggest a Semitic connection, if not an obviously Jewish one. If the Lemba are somehow Jewish now, how long have they been so? Are they simply a "Judaizing" tribe that adopted the religion relatively recently, or are they more likely descended from the ancient Israelites, as they say they are?

The Lemba claim to come from a place called Sena, somewhere north of their current African home. But until Tudor Parfitt undertook the role of anthropological sleuth in the 1980s and 1990s, no one knew exactly where Sena was. Some of the Lemba Parfitt spoke to used the term *Sena* in a more metaphorical sense—as a reference to heaven or the afterlife.

In concrete terms, Lemba tradition states that some twenty-five hundred years ago a group of Jews left Judea and, under the leadership of the Buba clan (one of the subclans the tribe was divided into), settled somewhere in the Middle East and built the city of Sena ("Sena One"). Conditions eventually became unfavorable, and the group left by boat. After landing on the African coast, the tribe split into two, with one faction settling in Ethiopia and another migrating farther south along the eastern coast of Africa. There, the second group is said to have rebuilt Sena ("Sena Two") somewhere in what is now Tanzania or Kenya. Another group then splintered from the tribe and built Sena Three in Mozambique. From there, it ostensibly went to Zimbabwe and participated in the construction of Great Zimbabwe, a stone city thought to have been built over the period extending from 400 C.E. to 1500 C.E. and covering nearly eighteen hundred acres—it is Africa's largest and most famous ruin. At some point, Lemba tradition goes, the Lemba people broke God's law by eating mice (definitely not kosher) and were banished from Great Zimbabwe and dispersed throughout southern Africa.

What Parfitt did was to attempt to trace the presumptive Lemba migration backward. Starting from the most densely populated Lemba enclaves in northeastern South Africa, he traveled north to Zimbabwe and Great Zimbabwe. The Lemba are particularly insistent that they had a hand in building Great Zimbabwe—again, this can never be proved, but in Parfitt's view it is plausible. Archaeological evidence demonstrates that Great Zimbabwe was constructed with wealth derived from trade and cattle. For centuries, the Lemba were known as major traders throughout southern Africa.

Proceeding farther north, Parfitt located Sena Three on the Zambezi River in Mozambique, although a civil war raging at the time prevented him from reaching the city. Given speculation on the part of some Lemba that Sena One could be found in Yemen, Parfitt continued across the Gulf of Aden, thinking that perhaps the town of San'a (or Sanaa, as it is sometimes written) may have been it. In the holy city of Terim, an imam informed him of a place called Sena at the end of the Wadi Masilah (Masilah valley) in central Yemen.

We still can't be completely sure that this Sena is Sena One, but the circumstantial evidence amassed by Parfitt seems to fit. First, to get to the sea, the Lemba would have had to cross the Masilah valley —perhaps this was Pusela. Second, a number of tribal names from that region of Yemen, the eastern Hadramaut, are extremely similar if not identical to those of the Lemba: Sadiki, Hamisi, Hamandishi, and others. Finally, Yemen's Sena was once supported by a massive stone dam. It is thought that about a thousand years ago the dam cracked and prompted a huge migration from the city. Was the break in the dam the unfavorable event that drove the Lemba to Africa?

Parfitt's remarkable journey and the presence of some elements of Semitic culture in Lemba culture should not allow us to get ahead of ourselves. The fact was, Lemba oral tradition was thoroughly at odds with the tribe's current geographic location and the language its members speak, to say nothing of their physical appearance. There was nothing in Parfitt's account, moreover, that ruled out an Islamic origin of the Lemba. Some scholars took issue with the Lemba narrative as well. A 1997 paper by a senior ethnographer in Harare, for example, claimed that the Lemba are purely African—the notion of a migratory tribe from the Middle East, he maintained, is purely an invention of outsiders. And indeed, as Parfitt writes in *The Lost Tribes of Israel*, such migration is exactly what has been ascribed to other tribes all over Africa.

The difference between the Lemba and those other tribes is that Lemba oral tradition posed a question that is readily amenable to genetic analysis because of the way tribal membership was (and is) regulated. Lemba traditions allowed a non-Lemba woman to join the community upon marriage to a Lemba man, but only after grueling initiation ceremonies. Some involved the woman's being set alight and forced to crawl through a hole in an anthill; others required her taking an emetic with sacrificial meat in order to vomit up her impurities. Such conversion was not open to non-Lemba men. Thus, non-Lemba DNA could enter the community only through females. In this way, Lemba practice inverts the typical Diasporan Jewish situation in which the mother determines ethnicity.

But the Lemba offered an ideal situation for study by my genetics research group. The Y chromosome passes exclusively down the male line. The Lemba claim that, historically, conversion has not been open to non-Lemba men, implying strict patrilineal descent of "Lembaness." If there is substance to the oral history, then Lemba Y chromosomes should look like the Y chromosomes of Jewish communities and not like the Y chromosomes of the Lemba's black African neighbors.

The first hint of something interesting in Lemba genetics came from the South African Trefor Jenkins, a world-renowned authority on the genetics of African populations. In 1996, Jenkins and his colleagues discovered that, despite the outward appearance and language of the Lemba, a significant fraction of their Y chromosomes was of apparent Semitic origin and not related to those of other Bantu speakers. In a study of some fifty Lemba men, Jenkins found that approximately

50 percent of the Lemba Y chromosomes were of Semitic origin, whereas 40 percent were so-called Negroid and the remaining 10 percent could not be assigned to any particular group.

Although Jenkins's results were intriguing, the most that could be said about them was that they were "consistent with the oral tradition of the Lemba." At that time, geneticists could not distinguish between Semitic and Jewish origins any more than anthropological observations could.

Not long after the Jenkins study, however, scientists discovered a slew of new genetic markers on the Y chromosome—this was during the steep growth curve of the Human Genome Project in the late 1990s. Markers began springing up everywhere, and even the barren tundra of the Y was no exception. These markers allowed much finer resolution of Y-chromosome relationships than previously possible. In particular, our study on the Cohanim made use of a number of these higher-resolution genetic markers and identified a chromosomal type that we thought might be a signature of the ancient Israelites. The Cohen Modal Haplotype, or CMH, is present in about half of the Cohanim tested and carried by more than one in ten Ashkenazi, Iraqi, and Yemeni Jews. With only one exception known so far (an Ethiopian population), this haplotype has not been found at such high frequencies in any non-Jewish group.

But even given better DNA markers and Jenkins's tantalizing results, I would be lying if I said we undertook this study without some apprehension. After all, it was still not much to go on. At the time we knew little about the genetic characteristics of the ancient Israelites. Our data strongly suggested that a large proportion of their Y chromosomes included the Cohen Modal Haplotype. But we also knew those ancient chromosomes would have included much more than just this haplotype. And we knew that the Lemba-Jewish connection, if there was one at the genetic level, could be to a Diasporan Jewish population wholly different from the ancient Israelites. Since the CMH is present, overall, in only one of every ten contemporary Jewish men, there was no guarantee it would have got into the Lemba via this route. And suppose the Lemba were Jewish but none carried the Cohen signature chromosome—what might our data look like in that case?

It all added up to a long shot. Still, Neil was as keen as ever and, having a bit of the gambler in my genes, I was too. I took a small sum of discretionary money (none of my Jewish genetics work has ever had any serious external funding) and rolled the dice.

To test the oral tradition surrounding the Lemba, we would need a large set of samples from multiple populations, and many of these samples would present particular challenges to collect. First, we would need samples from the Lemba, ideally from several areas. And to test whether Y-chromosome types found in the Lemba were unique to the Lemba, we would also require samples of local non-Lemba Bantu speakers. Finally, to test the idea of passage through Yemen, we would need Yemeni samples, ideally from the presumptive Sena One. Because of his strong personal contacts with members of the Lemba community, Parfitt was able to collect samples from two different locations. He also eventually arranged a sampling trip to Yemen for Neil and him, with the permission and armed protection of the government there. The overwhelming majority of Lemba asked to participate did so enthusiastically.

Thanks to Neil, Parfitt, and the Lemba, Mark Thomas and the rest of the wet-lab folks were able to type marker after marker and generate plenty of clean data. On the fateful day, Neil came over to my house in north Oxford bearing a huge stack of pie charts that showed the frequencies of all of the haplotypes in the various groups. We spread them out on my living room floor and had at it. So what did we find?

First, we had a look at the populations with which we were going to compare the Lemba. The Y chromosomes of the Israelites and individuals from the Yemeni Hadramaut were, in the main, very similar.

Both were very different from those of the non-Lemba South African Bantu speakers. The similarity of Hadramis (from Yemen) and Israelites was not surprising, since they are both Semitic groups, at least in their paternal heritage. Nor was it surprising that as a group the Semites were readily distinguishable from the more distantly related Bantu speakers. But the similarity of the Hadramis and the Israelites had one crucial exception. The CMH, common among the Israelites, was virtually absent in the Hadramis. Among a sample of a hundred individuals, only a single CMH was observed, and among the Bantu speakers, we saw absolutely none. This was good news from an interpretation perspective: the absence of the CMH from the otherwise related Hadramis meant that if CMH was observed in the Lemba the genetic contributions of non-Jewish Yemeni and Jews from virtually anywhere could be readily distinguished.

Because of the tendency of the Y chromosomes of Bantu speakers and Semites to differ, it was possible to sort through the Lemba Y chromosomes one by one and classify them as either Bantu or Semitic in origin. From this type of partitioning, we had a clear indication that the Lemba are different from their neighbors in more than just their stories. What it told us was that up to two-thirds of Lemba Y chromosomes come not from Bantu speakers but from a Semitic source, a conclusion presaged by the work of Trefor Jenkins.

But the real shocker came next: nearly one in ten of the Lemba Y chromosomes turned out to be none other than the CMH. Not only did we have evidence for a Semitic origin, we had an indication of a specifically Jewish connection. This seemed to be just what we were after, and Neil and I were agog that night.

And there was more to come. The Lemba have many different clans, and although Semitic Y chromosomes were spread throughout the clans, the vast majority of the CMH chromosomes observed were present in only one, the Buba. Moreover, the CMH was found at high frequencies (over 50 percent) in Buba from distinct geographic areas.

This was significant because of the Buba's unique role in Lemba oral tradition. Among Parfitt's papers was a carbon record of a funeral oration given by a tribal elder sometime before we started our research. It read in part, "The Lemba left Judea under the leadership of Buba and settled in Yemen, where they built their city of Sena." This record complemented a 1992 account by Matshaya Mathiva: "The Buba lineage came down from Judea as the leading lineage of the Basena ["people of Sena"], when they left Judea in their early migration to Yemen, where they settled and built the city of Sena."

Again, I would offer a caveat: no amount of DNA analysis can prove the claims of Lemba oral tradition. But after examining the Lemba's oral history together with the genetic evidence, I regard the most likely origin of the Semitic Y chromosomes among the Lemba as a Jewish one. The CMH is certainly not unique to the Jews. It is found once or twice in a hundred or so individuals in non-Jewish populations (less than about a tenth of its prevalence in Jewish populations). But combining the rarity of that chromosome among non-Jews with the oral tradition of the Lemba makes a Jewish origin seem likely.

Are there other possible explanations? Given the similarity of Jewish and Arab Y chromosomes, maybe the Semitic Y chromosomes in the Lemba were brought in by Arab traders at some time in the past. This seems unlikely to me, as my group was unable to find the CMH in high frequency in Arab populations: both the Palestinians and the Yemeni have very few.

Intriguingly, there is one non-Jewish group known to harbor high frequencies of the CMH: Ethiopians from Shewa and Wollo provinces. With the help of Ayele Tarekegn, an Ethiopian of Jewish ancestry who is a long-term collaborator of ours, we collected samples from this population. What we were able to obtain was limited to Tarekegn's social circle in the capital city of Addis Ababa in the province of Wollo. As a consequence, this collection may not be truly representative of the area. Other Y-chromosome data (not including the CMH) from genetic anthropologist Gérard Lucotte's lab did not find much in common between the Beta Israel and other Jewish groups.[5] Nor did a later study by Marc Feldman and Peter Oefner—the Y chromosomes of Ethiopian Jews looked much more like those of other Africans than they did those of other Jewish populations. Based on X-chromosome data, Damian Labuda of the Université de Montréal and his team suggested in 2005 that Ethiopian Jews might be of ancient Jewish origin; by Labuda's own admission, the evidence remains equivocal. Consistent with this inconsistent picture, work by Mike Hammer and his colleagues showed that some non-Jewish Ethiopians might have closer genetic connections to Jewish populations (either directly or indirectly) than Ethiopians who identify themselves as Jews. Thus, given what we know, it seems likely that the Ethiopian Jews (the Beta Israel) are descended from Ethiopians and converted to Judaism subsequent to any settlement of Jews in Africa. But the area that is now Ethiopia has a fascinating though poorly known history, with many potential connections to Jewish populations, and one gets the sense that there are many stories there waiting to be uncovered.

But let's suppose for a moment that the Tarekegn data are borne out. Could the Lemba CMH chromosomes have some link to the peoples of Ethiopia? Circumstantial evidence makes that possibility difficult to rule out: Semitic peoples appear to have entered Ethiopia in some number from the sixth century B.C.E. onward, whereas Amharic, the principal language of Ethiopia, is a Semitic language. Sena was long believed by some Lemba to have been in Ethiopia. Knives used for ritual slaughter by the Lemba are similar to those found in Ethiopia. Both the Lemba and the Beta Israel came to be known as prominent metal workers and stone builders. The two groups shared (and continue to share) similar patterns of endogamy, food taboos, and purity and burial rites. Perhaps some of these Semitic Ethiopians journeyed farther south and intermingled with peoples of the Bantu expansion.

An Ethiopian explanation for the Semitic chromosomes in the Lemba remains worth exploring. But in my view, it fails to account for the place accorded the Buba in Lemba oral tradition and the concentration of the CMH within this clan. Thus, we appear to be left with our original explanation: the Lemba Cohen Modal Haplotype chromosomes have a specifically Jewish origin.

Genetically speaking, how did the Lemba get to where they are today? Their history appears to involve a combination of Bantu and Jewish elements, perhaps augmented by other, non-Jewish Semitic ones. In genetic parlance, these types of combinations of different gene pools are known as admixture. We also carried out a different sort of genetic analysis demonstrating that the admixture event between Semitic and Bantu elements in the history of the Lemba happened fairly recently. So, might a white Cohen have introduced the CMH when European Jews began to arrive in South Africa in large numbers during the late nineteenth century? I don't think so. The Lemba's strict rules regarding endogamy and the fact that the CMH was observed in multiple, geographically distinct Buba communities would suggest otherwise.

When Trefor Jenkins's paper appeared in 1996, the Lemba took it as an affirmation of their Jewish identity. But even then, neither South African nor other Diasporan Jewish groups paid a great deal of attention.

For better or worse, that was not the case with our work. In fact, I probably helped the public relations machine along by opening my big mouth. When we were about to submit the paper to the *American Journal of Human Genetics* I gave a talk describing the work at the prestigious Cold Spring Harbor Laboratory on Long Island. Questioned afterward by a reporter, I said that I couldn't discuss the work at length because I would eventually submit an account to a journal (or at least that is what I intended to say). It is the general policy of scientific journals to permit scientists, before publication, to talk about their work in meetings but not with reporters. Certainly one is not supposed to tell a reporter what journal the work will come out in before the journal has had the opportunity to express an opinion—that's a foolproof recipe for egg on one's face.

Imagine my horror, then, when I picked up the *New York Times* one groggy Sunday in May 1999 to find a story on the Lemba on page 1, flanked by a picture of Matshaya Mathiva, with one of me inside the newspaper. "Dr. Goldstein said his findings had been submitted to the *American Journal of Human Genetics*," it read. I just about choked on my coffee. I quickly submitted the paper to them, thinking that if this thing gets rejected, oh boy, I am really up a creek. Thankfully, the paper sailed through the review process.

After the *Times* story was picked up by the rest of the media, the steady stream of reports began to have a profound effect on people's perceptions of the Lemba, to say nothing of the Lemba's perceptions of themselves. Despite public statements of caution from us, the discovery of the Cohen Modal Haplotype at such a high frequency among the Lemba was viewed as authoritative—it was the Cohanim all over again. Some Jewish groups in the United States, in particular, took the matter to heart and regarded it as proof positive that the Lemba were Jewish and should be accorded full rights of return to Israel. Hundreds of religious books and objects were sent to the Lemba from all over the world, while several "missionaries" were dispatched to bring the Lemba into the fold of "normative Judaism." By and large the Lemba respectfully declined such offers, preferring their syncretic Judeo-Islamo-Christian existence. Meanwhile, a number of other black African groups claiming Jewish roots became aware of what we were up to, viewing it as a vindication of sorts for their own oral traditions. Years later, I still receive the occasional e-mail from a tribesman wondering if DNA can help resolve some question of ethnic identity or other local conflict, Jewish or otherwise. Some are accompanied by interesting and sometimes tragic stories, for which DNA is unfortunately neither relevant nor helpful.

Personally, the Lemba project left me with a feeling of ambivalence. I am a geneticist with an abiding interest in human origins, histories, and migrations. I was heartened to learn that our findings had strengthened the Lemba's sense of Jewish identity and perhaps helped to validate their oral tradition. Of course, that was hardly my agenda—my lab was interested only in what the data might say. So what would have happened had we failed to find the CMH or any other presumptively Jewish genetic signature among the Lemba? How would they have reacted? And knowing what we know, what will our findings mean for the Lemba? Remember, they share genetic origins with other Jews based strictly on their paternal inheritance, whereas Judaism is traditionally defined by matrilineal descent. On that basis would they ever be accorded the same status as the Beta Israel—that of full-fledged Jews—by the Israeli rabbinate? And given the Lemba's syncretistic leanings, would they even want such status?

In retrospect, I've concluded that it is just not possible to control how my work is presented. Such are the pitfalls of practicing science so intimately connected to human populations and their sense of identity. All I can do is echo what Tudor Parfitt has written: "Jewishness is not a matter of DNA." Our genes needn't be our destiny, nor should they be.

As for the Lemba, when I take a step back, our results do not seem so shocking. After all, both cultural practices and archaeological evidence demonstrate just how far the Israelites traveled from their homeland in the hill country of Judea over the last three thousand–odd years. Jews have found their way to India and followed the Silk Road to China. Descendants of the Crypto-Jews of the Iberian Peninsula have been identified in the Balearic Islands off the coast of Spain, the West Indies and South America.[6] There is no reason to assume

Jews could not have reached the southern tip of Africa, even before the large-scale Jewish emigration from northern Europe to South Africa in the nineteenth and twentieth centuries.

Of course, accepting that version of events does not make the Lemba a Lost Tribe. But neither does it make their story any less compelling. As bioethicist Laurie Zoloth has observed, the story of the Lemba speaks to a central theme in the Jewish narrative: the repair of the Exile. Sixty years after six million died in the Holocaust, and in the wake of resurgent anti-Semitism in Europe and elsewhere, the prospect of finding new Jews in unlikely places has an undeniable resonance. To the extent that my colleagues and my work provoked a broader discussion of group identity—Jewish and non-Jewish—and complements our cultural understanding of ourselves, even in some slight way, I am gratified.

CHAPTER 3

Looking Out for Number Two:
The Case of the Ashkenazi Levites

And the Levites that are gone away
far from Me, when Israel went astray,
which went astray away from Me after
their idols; they shall even bear their
iniquity. Yet they shall be ministers
in My sanctuary, having charge at the
gates of the house, and ministering to
the house: they shall slay the burnt
offering and the sacrifice for the
people, and they shall stand before
them to minister unto them.
Ezekiel 44:10–11

In chapter 1, I tried to show that today's Cohanim are
paternal-line descendants of a community that effectively
maintained its genetic exclusivity over a period of at least

two thousand years. But the Cohanim were only a small subset of a larger priestly class, the Levites. Aaron was not only the first Cohen but also a member of the tribe of Levi and the great-grandson of the tribe's founder, Levi, the third son of Jacob. The tribe of Levi was distinct: it was the only one designated a priestly caste, it was not allotted a fixed territory of its own within the land of Canaan, and its members were counted separately in the census. Historically, the assumption has been that the Levites' priestly status was passed down from father to son, like that of the Cohanim. If that assumption is true, then it should be possible to identify a Y-chromosome signature in the Levites, just as Neil Bradman and Karl Skorecki rightly predicted we could in the Cohanim.

The Levites, however, would pose a challenge my colleagues and I had not faced with the Cohanim. The Levites' greater numbers (at the time) and their fate as Wandering Jews implied that they were a more heterogeneous bunch. In part to overcome that heterogeneity and in part because they were most interesting to us on a genetic level, we focused on the Levites who helped give rise to the Ashkenazi community. Might there be an Ashkenazi Levite–specific signature equivalent to the Cohen Modal Haplotype?

Without giving away the ending, I will tell you that what we found led us in an entirely unexpected direction. I can also say at this early stage that the genetics of the Levites are nothing like that of the Cohanim and that in fact they may even provide some support for one of the more romantic and, I would have thought, fanciful notions concerning the history of Ashkenazi Jews.

We know that the Levites were never accorded all the rights and responsibilities given the Cohanim. They were, as Risto Nurmela describes them in *The Levites*, a "second-class priesthood." The Cohanim, for example, got the first chance to read from the Torah; the Levites were to read second. Temple-sacrifice privileges—at the very core of Jewish sacred ritual—were limited to the sons of Zadok, whereas Levites became responsible for more mundane tasks such as day-to-day up-keep of the Temple, guarding the door, and transporting the Ark of the Covenant.[1] Was this consolation prize a historical accident? Was it punishment for Levi's crimes against the Canaanites or for Aaron's role in fashioning the golden calf (overall Levite faithfulness notwithstand-ing)?[2] Or was it for some other apostasy or iniquity, such as the one mentioned by Ezekiel in the epigraph above? We will never know.

In any case, by the time of the Second Temple, the non-Cohen Levites clearly had less influence and prestige than the Cohanim. As Nurmela notes, three sources—Ezekiel, the Priestly Code (P), and the Chronistic history (the companion volume to P)—are unanimous in their portrayal of the Levites as second-class priests.[3] The Levite position was therefore ambiguous: of high status but impoverished, "priestly" but not high priests.

It wasn't all drudgery, however. Although the Levites remained subordinate to the Cohanim, Levite responsibilities grew to include composition and performance of liturgical music, as well as teaching, administration, and interpretation of Jewish law and ritual. These func-tions became especially important during the Hellenistic period and, long before the destruction of the Second Temple, helped steer Jewish worship away from the Temple and into homes and synagogues.

Today the tribe of Levi, consisting of Cohanim and Levites, is the only one of the Israelite tribes not to have been lost, either figuratively or literally. Similar to the modern-day Cohanim, non-Cohen Levites are estimated to account for some 4 percent of the contemporary Ashkenazi community and somewhat less of the Sephardi population. They are found in synagogues all over the world; in more observant congregations, Levites continue to discharge their historical religious functions. They are called up second (after Cohanim) to read passages from the Torah. And while they no longer have occasion to enforce rectitude at sword point as they once did, in some congregations they still wash the hands of the Cohanim before the contemporary "priests" deliver their blessings. Neil Bradman, a Levite as far as he knows, often complains rather melodramatically that his lot in life will forever be hand washing.

Still, even today being number two is not without its privileges. Unlike Israelites (Jews not of Cohen or Levitic lineage), Levites do not have to pay a Cohen the ritual tax of five silver shekels on the birth of a first son.[4] And because he is not a priest, a modern Levite is also free of certain restrictions a Cohen faces. A Levite can, for example, marry a convert or a divorcee; he can also enter a cemetery. Such restrictions on Cohanim are enforced in the modern state of Israel, where religious doctrine influences some aspects of law (for example, marital law).

Historical position in the priestly hierarchy (for example, Levite versus Cohen) or in Jewish society at large (for example, Levite versus Israelite) is not the only way to classify a heterogeneous ethnic group such as the Jews. Geography is another. This approach has a direct bearing on the genetic origins of at least one group of Levites.

The two most numerous—and, prior to the explosion of Zionist resettlement in twentieth-century Palestine, historically separated— Jewish communities are the Ashkenazim and Sephardim. The term *Sephardi* originally described Jews descended from the communities that existed in Spain prior to the expulsion of all Jews in 1492 C.E. Now it is sometimes also used of descendants of the communities of North Africa and the Near East who follow the Sephardi rite of worship and cultural traditions.

The geographic origins and movements of Ashkenazi Jews are less well known. The term *Ashkenaz* appears in the Bible on several occasions and seems to refer to both a land and a people found somewhere close to the upper Euphrates and present-day Armenia.[5] Hundreds of years later (though probably no later than the sixth century C.E.), it came to be applied to the area populated by Jews in northern Germany between the Meuse and the Rhine. From the tenth century, Ashkenazi Jews spoke a common language (Yiddish) written with

Hebrew characters but with a lexicon borrowed mostly from German. By the sixteenth century, Jews speaking this language and following the Ashkenazi religious rites and cultural tradition populated communities extending from the Loire Valley in France in the west to the Dnieper River flowing through Russia, Ukraine, and Belarus in the east and from Rome in the south to the Danish border in the north. The Ashkenazi population continued to grow during the subsequent five centuries and is believed to have peaked at approximately thirteen million just prior to the outbreak of World War II.

Our studies of the Cohanim established that present-day Ashkenazi and Sephardi Cohanim are more genetically similar to one another than they are to either Israelites or non-Jews. Those studies also gave us an inkling that the Y chromosomes of the Ashkenazi Levites are different from those of the Cohanim, the Israelites, and even the Sephardi Levites. What they did not do was reveal a clear Levite-specific genetic signature comparable to the Cohen Modal Haplotype.

In fact, when we looked closely at the data, at the haplogroup level (defined by the combinations of unique mutations that tend to travel together), the Levites' diversity was what really stood out: the Levites turned out to have as many different Y chromosome types as the Israelites. Even more striking was that what we see in the Levites is not the same as or a subset of what we see in the Israelites. Among the Cohanim we see greatly reduced diversity, and what is present in Cohanim Y chromosomes is a clear subset of what is seen among the Israelites. The Ashkenazi Levites, on the other hand, have a very high frequency of a particular collection of Y-chromosome markers (a haplogroup called "R-M17") that is rare in Jewish populations and generally rare or absent altogether in populations of the Near East. Where did these Y chromosomes come from?

Once we had ruled out the possibility of a Cohen Modal Haplotype–like signature among the Levites, determining the origin of the mysterious Ashkenazi-Levite Y became the focus of our next study.

The marker patterns we saw along the R-M17 Y chromosomes told us they were unlikely to have come from the ancestral Hebrew population—they simply did not resemble other Y chromosomes from the Jewish populations we studied. But once we determined that they were completely different, we were stymied. How had such "foreign" chromosomes made their way into the Ashkenazi Levites but into no other Jewish groups?

Every detective knows the first place you look for a culprit is close to home. In looking for the origin of the Levite Y chromosomes, we started with the ancient neighbors of the Levites. We already knew of some populations that had a lot of R-M17 floating around. Work by my group and others had shown R-M17 to be prevalent in Scandinavia, eastern Europe, and western Asia, as well as in a geographic band south of Pakistan. In theory, the R-M17 chromosomes found in the Levites could have come from any of these populations.

Given those possibilities, could we make some educated guesses about which populations might be sources of Levite chromosomes? Not much is known about the origins or movements of the Ashkenazim. We do know, however, that European Jewish migration ultimately pushed eastward into Poland, Lithuania, and Russia. The Ashkenazi Jews therefore spent long centuries living among—and between—two great Eurasian tribal groupings: German-speaking peoples to the west and Slavic speakers farther east. Thus, in looking for the source of the Levites' haplogroup, we first considered the non-Jewish populations of northern Germany and Belarus. We chose northern Germans because they inhabit an area—perhaps *the* area—in which Ashkenazi Jewry first expanded. We selected Belarusians because they represent a significant number of the non-Jewish inhabitants of the Pale of Settlement extending from Lithuania in the north to Odessa on the Black Sea in the south.[6] Minsk, for example, the capital of Belarus, was once home to as many as fifty thousand Jews.

In addition to the Belarusians, we wanted a second Slavic group so that we could evaluate Y-chromosome differences from both eastern and western Slavic populations. The Slavic-speaking Sorbs (not to be confused with Serbs), now numbering about sixty thousand, are the only long-term linguistic minority in Germany, that is to say, near the presumptive western edge of Ashkenazi expansion. The Sorbs are concentrated in an area known as Lusatia in the eastern corner of Germany that borders what is now the Czech Republic.

Sorbs (also known as Wends) can be viewed as a Slavic holdout that survived west of the Germanic-Slavic boundary established by the rapid eastern expansion of the Germanic peoples in the twelfth and thirteenth centuries. But holding out hasn't been easy. Like their Polish and Baltic neighbors to the east, the Sorbs have been conquered repeatedly: by the Germans in 928, the Poles in 1002, and the Germans again in 1033. During the Nazi era, in an effort to make the Sorbs more Aryan, the Third Reich closed Sorb schools and churches and banned the Sorbian language. "It's a miracle that we still exist," the head of the Sorb Institute in Bautzen told *Time* magazine in 2005.

Geographic issues aside, we chose both Sorbs and Belarusians as Slavic groups in part because of the theories of Paul Wexler of Tel Aviv University about the origins of Yiddish (*Two-Tiered Relexification in Yiddish*), the language of Ashkenazi Jewry. Contrary to the almost universal view of Yiddish as a Germanic language, Wexler proposes that it was originally a combination of Sorbian and a pre-Ukrainian/pre-Belarusian (or, as Wexler calls it, "Kiev-Polessian") language that has since been "relexified" with German. (Relexification is the process whereby new vocabulary is substituted for old.) In other words, according to Wexler, Yiddish has a Slavic grammar and syntax, but those elements have been obscured because most of the original Slavic vocabulary has been replaced with German words.

Sometimes a population's taking on the language of another group is accompanied by sufficiently close social contacts that genes move between them. Thus, we thought the Sorbs and/or the Belarusians might be a source for Levite Y-chromosome haplogroups not only by virtue of their location and history but also because of their language.

In addition to these groups, we wanted to include one other population that had at least some R-M17 Y chromosomes but that wasn't considered a possible source for the origins of the Levite Y chromosomes. This population would serve as a sort of control group. That way if there was "incidental" sharing of types with the large grouping of R-M17 chromosome types, we'd see it in the control group as well as the test groups. For this purpose we chose Norwegians, who were known to carry a respectable frequency of R-M17 Y chromosomes.

Norwegians also represented an ideal control in light of a sad historical fact: Norway excluded Jewish immigration for the entire period during which the Ashkenazi community was forming and expanding —Jews have only been permitted to settle there since 1851. At the beginning of the twentieth century, the entire Norwegian Jewish population numbered 642. Today it is still less than 1,500. While studying genetic signatures of the Vikings, and with the BBC in tow, I had earlier embarked on a trip to collect Norwegian blood samples. I found the Norwegians I met polite and gracious though perhaps predictably not overly demonstrative. In a clumsy attempt at making conversation, I asked one of those donating blood for the project whether he could imagine his ancestors staging raids on the British Isles. "Oh yes," he assured me, "that was us."

Contemporary Slavs and Israelites are not very similar in terms of their Y chromosomes. Among the Israelites we typed, less than 5 percent of the Y chromosomes contained R-M17. By contrast, in the Belarusians and the Sorbs, the R-M17 frequencies were 50 and 66 percent, respectively. Among the Levites, 56 percent of the Y chromosomes were R-M17, a figure strikingly similar to that observed in the Slavic groups. Among Germans and Norwegians, the figures were lower (13 and 22 percent, respectively), but still much greater than what was seen among the Israelites.

Genetic and statistical analyses painted the same picture: only the Ashkenazi Levites, and not the rest of their fellow Jews, appear to have received a significant male contribution of Slavic origin. The Ashkenazi Levites were significantly different from all the populations considered with one exception: the Sorbs and the Ashkenazi Levites were statistically indistinguishable. That doesn't necessarily mean the Levites' Y chromosomes came from the Sorbs or that Yiddish is a Slavic language. It does mean that these chromosomes came into the Levites through a group that was similar to modern Sorbs in paternal genetic makeup.

But what happened when we added in more Y-chromosome markers to study the R-M17 chromosomes in greater detail? Using more extensive microsatellite variation, we found that among the Levites the Y chromosomes were mostly close relatives of one another. They formed a tight group of just three haplogroups, separated from one another by single microsatellite mutations. Furthermore, these three most common types were either completely absent or else barely present in the Sorbs and Belarusians. The case for Slavic origins of Ashkenazi Levites was showing some serious cracks.

When we started the Levite study, we intended to address the simple question of whether the Mishnah was correct: has paternally defined Levite status been faithfully transmitted as was observed for Cohen status? Our data say no and provide a negative control for the results of the Cohanim studies: we can find no convincing genetic signature of a high level of patrilineal inheritance among the broad community of self-identified Levites—their Y chromosome complement is much more heterogeneous than that of the Cohanim. Thus, the genetic evidence appears to favor the suppositions of John Bright (*A History of Israel*) and Risto Nurmela (*The Levites*) that it was much easier to become a Levite than it was a Cohen. To become a Cohen usually required a father who was a Cohen. To become a Levite probably required no more than faith and conviction (and perhaps the occasional well-placed bribe). Hence, we see all manner of Y chromosomes among Levites.

But our work also led to a new question involving the one Y-chromosome signature we did identify in great numbers, R-M17. Since R-M17 is not common among current Jewish groups or, presumably, among the ancient Hebrews, how did it come to be in such high frequency in the Ashkenazi Levites?

One possibility is that there was never an influx of R-M17 chromosomes into the Levites from a Slavic or genetically similar source in the first place. Rather, the elevated R-M17 frequencies in Ashkenazi Levites may be due to chance, or so-called genetic drift. Genetic drift is the consequence of small population sizes. The gene forms that one generation passes on to the next generation are not all the gene forms (or variants) that generation has. Some individuals reproduce, others don't; some reproduce more, some less. If the gene variant in question has no effect on survival or reproduction, whether it is passed on is a matter of chance. When population sizes are small this can result in large changes in the proportions of gene forms from one generation to the next. Because every gene form is equally likely to be passed on or not when there is no selection on it, the phenomenon of genetic drift can be likened to coin tossing. If you toss a fair coin ten times, the likelihood of a large departure from the expected 50 percent heads is very high. You could easily get something very different, such as only 30 percent heads. But if you toss it ten thousand times, the odds of such a radical departure from expectation are very small. The same is true in population genetics. When the population is small, large changes in the frequency of types of genes can occur by chance, leading to extreme increases or decreases in the frequencies of particular types. In the case of the Levites, the relevant population size may have been small enough to result in chance changes in haplotype composition.[7]

Genetic drift therefore must be considered as a possible explanation for the apparently foreign chromosome type common among the Levites. While this is a formal possibility, it seems unlikely to me. If the Levites were isolated from other Jews in their paternal heritage, we would expect to see overall strong differences such as in the Cohanim. Or if drift occurred in the broader Jewish community, then R-M17 should be in other Jews as well. But what we see is R-M17 at high frequency *only* among the Ashkenazi Levites.

Our data suggest that the R-M17 chromosomes appeared among the Ashkenazi Levites approximately a thousand years ago. Historically, such a date would appear to coincide with the consolidation and expansion of the Ashkenazi community. If the R-M17 chromosomes did not originate with the Sorbs' or the Belarusians' ancestors, where did they come from? Is there another candidate population with a putative Jewish connection?

Somewhere in central Asia, perhaps near the modern nation of Mongolia, more than two thousand years ago, a group or groups of nomadic tribes with a rigid and characteristic social organization developed exceptional skills as horsemen. The Turks became mercenaries for various Asian overlords and, following their masters throughout Asia, reached as far west as modern Bulgaria in the first centuries C.E. The overlords mostly faded from view, and the Turkish mercenaries assumed political control, establishing empires that at one time or another encompassed most parts of Asia (including, most famously, the Ottoman Empire).

Among the Turkic powers to emerge were the Khazars, who until 630 C.E. were part of the western Turkish Empire. (Appropriately enough, the name Khazar appears to derive from a Turkish word meaning "wandering.") After that empire's collapse, the Khazars established a Central Asian empire of their own along the Caspian Sea that converted to Judaism in the ninth century C.E. Although we do not know how much of the general population of Khazaria may have converted, correspondence between Khazaria and Jewish officials in Moorish Spain records Judaism as the state religion of Khazaria. From the tenth century on, the major Khazar Jewish documents were written in Hebrew. It's not clear why the Khazar King Bulan chose to convert his people to Judaism, but historians speculate the reason may have been political. To become Muslim would have meant accepting the leadership of Khazaria's archenemy, the caliph (the one-time temporal and spiritual leader of Islam). On the other hand, conversion to Christianity might have led to excessive dependence on Constantinople, the Christian stronghold of the Byzantine Empire.

In its time, Khazaria fought the Arabs to the south, maintained friendly relations with Byzantium, and corresponded with the Jewish chief minister of the Cordoba caliphate in Muslim Spain. Eventually, however, Khazaria succumbed to the expanding Slavic peoples. In 1016, Khazaria fell to the new Viking nation of Kievan Rus. Some of Khazaria's subjects are thought to have migrated west to central and northern Europe. As early as the tenth century, Duke Taksony of Hungary invited Khazars to live under his protection.

Some scholars and popularizers, most notably Arthur Koestler (1905–1983), have suggested that Ashkenazi Jewry may derive more from Khazari sources than from ancient Israelites. The Khazar Empire, they point out, was unarguably the most powerful Jewish polity during the long period after the revolts against Rome and before the establishment of the modern state of Israel. Koestler, the Hungarian-born journalist, novelist, and public intellectual, made the case for Khazari roots of the Ashkenazim in his controversial 1976 book, *The Thirteenth Tribe*.

The Khazar Empire stood to the east of the area of the Ashkenazi Jews, between the Dnieper and Volga rivers, and reached south into the modern nation of Georgia. Critically, although Khazaria was established by Central Asian Turkic-speaking peoples, it encompassed the Slavic heartland. The citizens of the empire, therefore, were most probably Slavic and no doubt rich in R-M17. Could Khazaria, I wonder to this day, be the source of Ashkenazi Levite R-M17 Y chromosomes?

As with much else of genetic history, there is no way to be sure. It would, however, be well worth investigating whether any documentary evidence exists of high-status individuals, noblemen, or priests of Khazaria, for example, adopting Jewish status or, more critically, Levite status. It would also be worth investigating further the genetic structure of the geographic regions that were formerly under Khazari control, both Jewish and non-Jewish. A number of historians have proposed that at least some of the Mountain Jews of Dagestan (a small community in the north Caucasus) are a direct legacy of Khazari Jewry. And in a 2005 paper, Marina Faerman and colleagues at Hebrew University found evidence of R-M17 having made its way into the Ashkenazi population right around the time when the Ashkenazi community was coalescing in Europe (Nebel et al., "Y Chromosome Evidence"). They, too, speculated that R-M17 might be a genetic legacy passed down from the Khazars.

And there are other avenues open for investigating the question of whether Khazarians are the source of Ashkenazi Levite R-M17 Y chromosomes. For example, the extant language most closely related to Khazari is Chuvash, spoken by more than a million citizens of the Chuvash Republic of the Russian Federation, located in central Russia. A careful comparison of their Y chromosomes with those of the Levites seems very much in order.

We started out asking a fairly simple question: Do the Y chromosomes of the Levites, like those of the Cohanim, show the signature of strict patrilineal inheritance? From this simple beginning, however, we somehow found ourselves navigating Arthur Koestler's fervid imagination. Toward the end of his life, his eclectic palette of passions came to include the idea that he and his fellow Ashkenazim must be genetically descended from the Khazars.

Koestler was no historian, and in a rare moment of humility he admitted the hypothesis was in part designed to undermine anti-Semitism by connecting the lineage of European Jews to that of Gentile Europe.[8] And despite his brilliance, Koestler was regarded by many as an amateur or even crank in some of his theorizing. His attacks on Darwin and defense of Lamarckian inheritance, allegations he mistreated women, and his supposed suicide pact with his apparently healthy last wife did little to disabuse people of that sort of impression.[9] For my own part, I must confess that, like virtually every academic I have ever consulted on the subject, I was initially quite dismissive of Koestler's identification of the Khazars as the "thirteenth tribe" and the origin of Ashkenazi Jewry. Was this not just another self-aggrandizing Lost Tribe narrative bereft of evidence?

I am no longer so sure. The Khazar connection seems no more far-fetched than the spectacular continuity of the Cohen line or the apparent presence of Jewish genetic signatures in a South African Bantu people. And then there is that troubling Y chromosome that is so common in the Ashkenazi Levites but seemingly nowhere else to be found. I cannot claim the evidence proves a Khazari connection. But it does raise the possibility, and I confess that, although I can not prove it yet, the idea does now seem to me plausible, if not likely.

CHAPTER 4

Those Jewish Mothers: The Development of Female-Defined Ethnicity in the Jewish Diaspora

The elders said, "We are witnesses.

May the Lord make the woman

who is coming into your home like

Rachel and Leah, both of whom

built the house of Israel."

Ruth 4:11

Up to this point, we've considered Jewish genetic history exclusively from the male point of view, that is to say, the Y chromosome. This orientation not only is not fair but also paints a distorted picture—there are clear differences between the male-exclusive Y and the rest of the human genome. From my perspective as a genetic historian, the Y chromosome is inadequate, because it says nothing about the patterns of migration among females. To fully understand the genetic history of a population it is necessary to move beyond the Y chromosome.

My interest in considering female Jewish genetic history is about something more than just fair play and clearer genetics. The fact is, for most of the history of the Diasporan Jewish experience it has been the mother who has determined religious identity: if your mom was Jewish, you were Jewish. This long-established custom has only recently been relaxed by some reform movements in the United States. As anthropologist Melvin Konner writes, the post–Second Temple rabbis made the Jewish mother the sine qua non of Jewishness. But given the prominence of matrilineal descent in Jewish tradition, we know surprisingly little about its origins. We have no indication that the mother determined ethnic or religious identity in the ancient Near East in general or among the ancient Israelites in particular. Indeed, there are many indications to the contrary. In the Bible, people are often named after male founders: Moabites after Moab, Ammonites after Ben-Ammi, and Edomites after Jacob's brother Esau. More directly, when Old Testament parents have different backgrounds, the children are usually identified as members of the father's ethnic group. Ruth, for example, is a Moabite who marries an Israelite of the Ephrathite clan, but when the Bible introduces her famous great-grandson, it says, "David was the son of an Ephrathite named Jesse." No question King David was an Israelite, despite having a "non-Jewish" great-grandmother whose conversion was no more than a simple declaration to her mother-in-law, Naomi, that "your people shall be my people, and your God my God" (Ruth 1:16). Similarly, Solomon's son Rehoboam had an Ammonite mother, but his status as an Israelite was never disputed. Even today among the Samaritans, who claim direct descent from the ancient Israelites, the father, not the mother, determines religious identity.[1] Despite this apparent history of paternal identification, sometime well after the Babylonian exile Jewish identity came to be determined by the mother and not the father.

We do not know exactly where or when this practice started. Both Mishnah Kiddushin (3:12) and Mishnah Yebamot (7:5), presumed to be written circa 80–120 C.E., lay out various criteria as to what constitutes a valid marriage and the status of the offspring of those couplings.[2] Subsequently, the Babylonian Talmud was clear: "Thy son by an Israelite woman is called thy son, and thy son by a heathen is not called thy son but her son" (Kiddushin 68b). From that point on (circa 550–700 C.E.), the custom was followed by all major branches of Judaism until the late twentieth century, when Jewish Reform, Reconstructionist, Humanist, and Renewal movements in North America relaxed or abolished the use of matrilineal descent as a criterion for Jewishness.

The notion of matrilineal descent is clearly not biblical if we use only the Hebrew Bible as a basis for information. Abraham, Judah, Joseph, Moses, and David all took foreign women as wives. And although it's hard to figure out how he managed it, Solomon married seven hundred Gentiles and took another three hundred as his concubines (1 Kings 11). As Shaye J. D. Cohen observes, nowhere is there any indication that these marriages were invalid, that the children produced by them were not Jewish, or that the women were expected to convert to Judaism.

So why the change of precedent in the Talmud? There are several possibilities. One has to do with the peculiarities of human reproduction. As the ancient Latin dictum puts it, *mater certa, pater incertus:* the identity of the father might not be knowable (at least prior to the advent of forensic DNA), but the identity of the mother always is. The rabbis may have felt that a child's identity could be ascribed with much more certainty by linking it to his or her mother.

Shaye J. D. Cohen finds more plausible two other explanations for the maternal transmission of Jewishness. One is that the introduction of matrilineal descent coincided with the rise to prominence of similar Roman laws governing intermarriage. If two Roman citizens married, for example, their children's status was determined by the father. But if a Roman citizen impregnated a female slave, the child would be a slave. In the case of the Jews, matrilineal descent may have been a response to a long history of persecution and the ongoing struggle to retain their religious and ethnic identity in the Diaspora. Female-defined ethnicity might also have developed in part to keep Jewish men from alienating their hosts by competing with them for wives.

Cohen's second hypothesis is that matrilineal descent arose organically from rabbinical thought on the subject of reproductive mixtures of different types—in genetics what we call "negative assortative mating." The authors of the Mishnah, he notes, were fairly obsessed with animal husbandry, and he cites examples considered to place primacy on the female animal in determining what happens with the offspring.

How to Study Female-Defined Ethnicity

Whatever the reasons for the development of female-defined ethnicity in Diasporan Jewish culture, faithful adherence to it would mean that the genetic heritage of Jewish populations is different for males and females. Our primary aim, therefore, was to use genetics to investigate this long-standing Jewish custom.

Ideally, if we want to look for a genetic signature of matrilineal descent, we'd like to have a female equivalent of the Y chromosome, one that is passed exclusively down the maternal line. So why not the X chromosome, which is so closely associated with being female? For a genetic historian, there are a couple of problems with the X. First, it is not at all female-specific: females have two X chromosomes and transmit one of them to each child. Fathers, in turn, give their daughters their only copy of the X chromosome. Second, unlike most of the Y, the female X undergoes the genetic reshuffling process that drives evolution with every conception. Thus, whereas the Y is transmitted nearly unchanged over many generations, the X is much more dynamic and therefore more difficult to track backwards through time. This is why, in order to assess female history, we turn to the mitochondria.

The mitochondria are discrete packages, or organelles, found outside the nucleus of cells. These organelles are essential to cells because they are the site of the final breakdown of food and the subsequent liberation of energy—in essence they are the power plants of the cell. The origin of mitochondria goes back more than a billion years, to a time before there were males and females—indeed to a time before there were plants and animals.

Mitochondria were originally free-living bacteria somewhat similar to rickettsia, the cause of typhus, an infectious disease transmitted by lice and fleas. Early in the history of life on earth, one group of cells absorbed mitochondria to their interior to the mutual benefit of the host cell and the mitochondria, an arrangement called symbiosis. In time, some of the genes necessary to the function of the mitochondria moved into the genome inside the nucleus of the mitochondria's host, and the mitochondria and the host became forever dependent upon one another. Though reliant on the largesse of its host cell, the mitochondrion retains a small genome of its own, the mitochondrial DNA.

Although the biology of the mitochondrion is fascinating and the subject of much research, I am less interested in its inner workings than I am in its inheritance pattern. Just as the Y chromosome is transmitted exclusively through the male line, mitochondrial DNA is inherited strictly through the female line, and although it is passed from mothers to both sons and daughters, sons do not pass on their mitochondrial DNA. Again, this is one of the peculiarities of mammalian reproduction. Both female eggs and male sperm have many mitochondria (as do most cells in the body). In the egg, mitochondria are widely distributed, while in the sperm they are concentrated around the tail, which is the motile force for the sperm and the site of energy production. When the head of the sperm penetrates the egg, the tail is discarded along with all the paternally supplied mitochondria.

The mitochondria in the fetus therefore come only from the mother. The male readers of this book and I carry our mothers' mitochondrial DNA, but for us and our brothers, the journey ends there. So, although the mechanism for mitochondrial DNA transmission is different from that for Y chromosome transmission, the end result—sex-specific inheritance—is the same. The Y chromosome records the history of males in a population, and the mitochondrial DNA records the history of females.

There are a couple of other things to know about mitochondrial DNA. Like the Y chromosome, when it passes from father to son, but unlike the rest of our nuclear genome (chromosomes 1 through 22, X and Y), the mitochondrial genome is passed from mother to daughter intact. Different mitochondrial DNA molecules do not usually mix with one another. This means that all the mitochondrial DNA on the planet, just like all the Y chromosomes on the planet, share a single evolutionary history. That is, there is a single tree for this DNA segment with a root genome, or mitochondrial Eve, leading to all current mitochondrial genomes, just as there is a single distinct tree for the Y chromosomes, with a Y-chromosome Adam at its root. Here the biblical analogy breaks down. Because of the vagaries of genetic history, our Y-chromosome Adam, the most recent common ancestor of the Y chromosome of all men alive today, lived around sixty thousand years ago, some eighty-five thousand years *after* mitochondrial Eve, the most recent common ancestor of every female lineage we see today. Thus, this particular Adam never met, let alone dated, this particular Eve.

The mitochondrial DNA molecule is very different from the Y chromosome. First, it is a very short stretch of DNA, weighing in at less than seventeen thousand base pairs, compared to sixty million for the Y chromosome. Second, unlike the mostly inert Y, the major part of the mitochondrial DNA molecule encodes genes or regulates their expression. That is, most of the DNA present in the mitochondria makes protein and the exact sequence must be maintained precisely in order to encode the correct sequence of amino acids that make the different proteins. For this reason, presumably, there is very little repetitive DNA in the mitochrondial DNA, or mtDNA, molecule (for example, it does not have rapidly evolving microsatellites, as does the Y chromosome and other parts of our genome).

Nevertheless, there are analogues of the unique mutations and microsatellites of the Y chromosome available in mitochondrial DNA, because some sites in the molecule evolve much faster than other sites. The so-called control region is one such site. It serves only to initiate replication of the mitochondrial DNA and the copying of mitochondrial genes involved in energy production—this part of the mitochondrial does not code for the proteins themselves. This region is therefore less constrained by evolution and, except for a small con-served region, has a much higher rate of change than other parts of the mitochondrial DNA genome, which is really saying something. The rest of this genome, while not changing as fast as the control region, still changes almost ten times faster than nuclear genes, although no one is quite sure why. In any case, depending on the study at hand, researchers may choose to use the relatively faster or more slowly evolving sites. In my lab's study of matrilineal descent, we focused on the rapidly evolving control region, which allowed us to assess more changes and make a finer-scale comparison of genetic diversity in different populations.

Mitochondrial DNA, like Y-chromosome DNA, can tell us about the history and migration of only one gender. Therefore, to really understand the effects, if any, of matrilineal descent on entire Jewish populations, mitochondrial DNA would not be enough—we would need to examine Y-chromosome data in multiple populations, too. If we did so, we would be better able to assess not only whether matrilineal descent was adhered to over any length of time but also whether ancient Jewish Diaspora communities were more likely founded by Jewish men and local (non-Jewish?) women (or vice versa) and whether those founding communities were genetically distinct in different places. In addition, examination of Y chromosomes would presumably tell us whether ancient Jewish communities were truly endogamous or if instead there was male input from the host populations. In other words, to understand the maternal history, we would need to know what the fathers were up to as well.

From our experience with the Cohanim, the Lemba, and the Levites, we already knew that a particular combination of Y-chromosome genetic markers would give us a high degree of resolution for comparing the types of Y chromosomes in Jewish and non-Jewish populations. In particular, we looked at thirteen unique mutations and six microsatellite polymorphisms. Recall from chapter 1 that the former are so named because they have occurred very rarely in human evolution. Consequently, the vast majority of chromosomes carrying a "unique" mutation are more closely related to one another than any of them are to chromosomes that do not carry the mutation. Thus, these unique mutations identify relatively large groups of related Y chromosomes. But sometimes we need to work out relationships of the Y chromosomes *within* these larger groups. For this purpose, genetic historians use the much more quickly evolving microsatellite markers—that, of course, is exactly what we did for our studies of the Cohanim and the Lemba described in chapters 1 and 2.

Once we resolved to look at the history of Jewish populations through the window of both Y-chromosome and mitochondrial DNA genetic variation, we had to be sure we had a valid way of analyzing our results. Why not compare the boys with the girls, that is, the Y chromosomes with the mitochondrial DNA? Unfortunately, it's not that simple. Comparing Y chromosomes and mitochondrial DNA is like comparing apples and oranges. Why?

One key difference is that the mitochondrial DNA has a higher mutation rate than the Y-chromosome DNA sequence (excluding the rapidly evolving microsatellites), although to what extent is not clear.

In any case, differences in mutation rate lead directly to different amounts of genetic diversity between Y chromosomes and mitochondrial DNA, even if there are no differences in movement patterns or other demographic factors between separate male and female populations. Unfortunately, we do not know enough about the mutational differences to properly account for them in direct comparisons between mitochondrial DNA and Y-chromosome variation.[3]

Beyond mutation rates, other differences complicate direct comparisons between the two genetic systems. Whatever effects female-defined ethnicity might have on Y-chromosome and mitochondrial DNA variation, these two parts of our genetic makeup will also be influenced by cultural practices that throughout history have been common among many populations, Jewish and non-Jewish. One such cultural practice is reproductive behavior. In many human societies, men have multiple sexual partners, either officially or unofficially. This practice can result in some men having a great many offspring while other men miss out entirely. For example, Emperor Moulay Ismail "the Bloodthirsty" of Morocco, who ruled from 1672 to 1727, is thought to have sired somewhere in the neighborhood of seven hundred sons and three hundred daughters, earning a place in *Guinness World Records*. Obviously, by comparison the number of children a woman can produce is strictly limited. The female record holder, a Russian peasant, produced sixteen pairs of twins, seven sets of triplets, and four sets of quadruplets—a total of sixty-nine children—between 1725 and 1765. For a woman to give birth to that many children is remarkable, to say the least, but it's still a small fraction of the emperor's brood. As a consequence of this type of sexual asymmetry, the chances of men and women contributing their genes to the next generation may be markedly different. Differences in the number of the sexes is one of the factors that can make a large population effectively behave more like a smaller one in terms of the consequences for genetic variation in the population. This idea is captured in a formal way in population genetic terms as an "effective population size," which is normally smaller than the real census size of a population.[4] For example, if we were to look at the Y chromosomes in a village that included male descendants of Moulay Ismail, we might find a preponderance of a single Y, that of Ismail himself. But if instead we examined the mitochondrial DNA of both male and female descendants of Ismail in that same village, we would see many different mitochondrial DNA types all contributed by various members of the emperor's harem of five hundred–plus women and, of course, none from the emperor himself.

Because direct differences between the evolution of a population's mitochondrial DNA and Y chromosomes can be due to any number of reasons, it would be reckless to ascribe differences solely to female-defined ethnicity. This complication, unfortunately, has not stopped some geneticists from making direct comparisons between Y chromosomes and mitochondrial DNA. But if we are truly interested in the history, we must admit that simply comparing Y chromosomes and mitochondrial DNA in Jewish populations is insufficient. What then should we do?

The alternative is to make direct comparisons within each of these very different genetic systems: compare mitochondrial DNA with mitochondrial DNA and Y with Y, or apples with apples and oranges with oranges. Seems logical enough—so why has this been done so rarely? Because it takes a lot of work! It requires a set of population samples collected specifically for the purpose of assessing the genetic impact of female-defined ethnicity. Although it may come as a surprise, sample sets are rarely collected specifically to allow historical inference —funding agencies such as the National Institutes of Health are concerned more with health than with history, and rightly so. That means that genetic history researchers try to do whatever can be done with the samples already available, most of which have been collected for medical genetic studies. To study the genetic effects of Jewish cultural practices properly, we would therefore need to undertake a massive collection effort. With the new samples, we would then make direct comparisons between mitochondrial DNA in Jewish populations and mitochondrial DNA in the host populations among whom they live. We would need to do the same for the Y chromosome. And that's exactly what we did.

Since we were about to enter the then-unfamiliar world (to us at least) of mitochondrial DNA, I thought we ought to get some help. I brought on Martin Richards, now a senior lecturer at the University of Leeds and a respected authority on mitochondrial DNA evolution. I'd met him at Oxford and was extremely impressed with his facility with mitochondrial DNA and his understanding of how it could be used in genetic history. Unfortunately, it became clear almost immediately that he and Neil rubbed each other the wrong way. Science is hard enough as it is—clashes of personalities can easily become a distraction and grind the work to a halt. Eventually, after providing a valuable intellectual foundation for the work, Martin moved on and the project was handed over to the ever-reliable Mark Thomas and Michael Weale.

There are many reasons I was convinced paired comparisons of Jewish and host populations would help uncover the genetic histories of Jewish populations. For example, one of my primary interests was in asking how much of the gene pool of a given Jewish population is derived from an ancestral Israelite population and how much is derived from the local host population due to religious conversion or other means of gene flow between the hosts and the Jews. Having both potential sources of genetic material increases the chances of being able to make this distinction. As usual, it is not quite so easy as it sounds. The ancient Israelites are not, of course, available for study, so our group had to use indirect approaches for assessing their genetic composition. And although recent host populations suitable for matching with Jewish populations were available, our uncertainty about Jewish history meant that identification of the most appropriate host populations amounted to a series of educated guesses.

So what about those populations? Before World War II (1939–1945) and the founding of the modern state of Israel (1948), many long-standing separate Jewish communities existed in Europe, North Africa, and Asia. For this study, our goal was to include representatives of as many of these groups as possible as well as the people among whom they lived. In total, we included eighteen populations from Europe, North Africa, the Middle East, Central Asia, and the Indian subcontinent. In each case, we paired a Jewish community with a neighboring non-Jewish population for comparison.[5] We also included a set of Israeli and Palestinian Arabs to represent possible long-term residents of the biblical land of Israel.[6]

Some groups were fairly easy to choose in light of history, language, and geography—Georgian Jews and non-Jewish Georgians, for example. Others took some thought and some digging into the histories of the Jewish populations and their respective regions. In every case the work was great fun—it taught me a great deal about Jewish cultures far and wide, many of which I did not even know existed.[7] To wit:

- The Berbers, an indigenous North African tribal people thought to have come to Morocco from the Iberian Peninsula around 1000 B.C.E., had a close relationship with the Jews of Morocco. When Arab armies invaded North Africa in the seventh century C.E., Jews and Berbers fought side by side. Indeed, under the leadership of a Jewish Berber queen of priestly descent known as El Kahina, the two peoples were able to hold off the Arabs for a short period. After leading the Jews and Berbers in a guerrilla war in the mountains, El Kahina was eventually captured and beheaded, thereby effectively ending dreams of Berber independence. Many if not most pagan and Judeo-Berber tribes were then forcibly converted to Islam.
- Before 1948, there were three bustling Jewish communities in India: the Baghdadi Jews of Mumbai (Bombay) and Calcutta, the Cochin

Jews along the Malabar Coast (the southwestern tip of India), and the Bene Israel, mainly in Mumbai. Until the Portuguese gained a foothold in coastal India in the sixteenth century (and brought the Inquisition with them), the Bene Israel was that rarest of Jewish communities, one that lived virtually free of religious persecution and largely in harmony with its Hindu and Muslim neighbors. Among some traditional Hindus, pressing oil from seeds was forbidden because it was considered destruction of life. By contrast, Jews had a centuries-old tradition of making olive oil. Thus, the Bene Israel found a lucrative niche and became known as the "Saturday Oil Presser Caste" on account of their skill at making various oils and their observance of the Jewish Sabbath. Many Bene Israel members reached the upper echelons of Indian society; when Mahatma Gandhi undertook his hunger strikes against the British, for example, a Bene Israel physician often tended him.

- In the summer of 1942, the Nazis marched into Azerbaijan intent on capturing the oilfields of Baku; for whatever reason, however, they never reached Georgia to the west. Consequently, the Georgian Jews were one of the very few Jewish communities to escape major losses during the Holocaust.

- Some Yemenis spend about one-third of their income on *qat,* a sour-tasting and mildly narcotic mountain shrub known as the "flower of paradise." Prior to the waves of emigration to Israel, many Yemenite Jews were among habitual *qat* users, sometimes chewing the plant in synagogue or during prayer or study sessions.

In doing my research on the Jews and their neighbors, I was struck by several themes that seemed to recur throughout Jewish history, no matter how remote. The first is the obvious motif of persecution. It hardly seems necessary to point out, but from Yemen to Georgia, from Germany to Ethiopia, the Jews have traveled a brutal path and repeatedly paid for their continued existence in blood. Whether at the hands of Christian or Muslim extremists, Fascists or Communists, anti-Semitism seems to be a deep and abiding global tradition. When Jewish songwriter and satirist (and Harvard-trained mathematician) Tom Lehrer sang "And everybody hates the Jews," he wasn't kidding. A notable exception is the subcontinent of India, which for whatever reason, never indulged in systematic anti-Semitism.

The second theme is one that was made apparent to me when we undertook the study of the Lemba: the notion of descent from the Lost Tribes. The Bene Israel and Bnei Menashe of India, the Beta Israel of Ethiopia, the Ebraeli of Georgia, the Bukharan Jews, and many others all cherish the idea that they can trace their lineage back to the Assyrian Exile of 722 B.C.E. or thereabouts.

Finally, there is the recurring theme of adjustment to a new homeland. Again and again I read about the difficulties non-Ashkenazi Jewish groups have faced in assimilating themselves into Israeli society while retaining their unique ethnic identities. The Bene Israel of India, for example, were not accorded the same rights and privileges as other Jews until an act of the Israeli parliament in the mid-1960s. To this day Jews from India settle disproportionately in Dimona, a development town in southern Israel that is known as the center of Israeli expertise in nuclear energy (and, by popular assumption, weaponry). Upon arrival, the Beta Israel of Ethiopia were often treated with indifference by the Israeli populace and suspicion by religious authorities. In one infamous incident, an Israeli newspaper discovered that for years the blood of Ethiopian donors had been discarded by Israeli medical officials for fear of HIV/AIDS. This disenfranchisement was also captured in a haunting popular song by Ethiopian immigrants lamenting that if their mother, lost in the long trek through Ethiopia, were there she could "tell them that I am Jewish." The Georgian Ebraeli (now known as Gruzinim), too, were regarded with suspicion by many Israelis for their clannish and iconoclastic ways; a small minority became part of a burgeoning Israeli underworld. Meanwhile, the newly immigrated Yemenites were often consigned to unskilled labor and treated as an inferior tribe. At a time when immigration is such a hotly debated topic in the United States, the accounts of the hardships of Jews wishing only to exercise the right of return to their ancestral homeland offer a vivid reminder of how much is at stake when whole populations relocate, even by choice.

These themes converge in the ongoing story of the Bnei Menashe ("children of Manasseh"). Members of the Kuki-Chin-Mizo tribes residing in northeastern India near the Myanmar border, they are said to number as many as seven thousand. Like the Lemba, for generations they have observed strikingly Jewish-like rituals such as circumcision and a festival marked by unleavened bread; many formally converted to Orthodoxy in the 1970s. Several hundred have migrated to Israel. Hillel Halkin began to follow the Bnei Menashe story as a deep skeptic; he and many others are now convinced that the Mizos are indeed remnants of the Lost Tribe of Manasseh. In 2002, Halkin agitated for the Mizos to undergo DNA tests in order to vindicate their claims. Most resisted. A recent paper in *Genome Biology* (Maity et al., "Tracking the Genetic Imprints") found no evidence for Cohanim chromosomes in the Bnei Menashe and only equivocal evidence of a Near Eastern origin on the maternal side.[8] Some commentators took the opportunity to say, essentially, "Ha—told you they weren't 'real' Jews." But in 2005, the Sephardic rabbinate recognized them as Jews anyway; by the time you read this it is likely that many will have made their way to Israel. In my view, this is a perfect example of when we are best served by ignoring DNA—if living, breathing people want to embrace Judaism, genetics should have nothing to say about it one way or the other. As one American rabbi succinctly and eloquently put it, "DNA Shmee NA."

So given our eighteen populations, the question we wanted to ask was whether they carry a genetic legacy of their relatively peculiar practice of defining ethnicity through the mother. We know that mitochondrial DNA and Y-chromosome variation reflect, in some way, female and male genetic history, but how should we use variation in these systems to assess the effects of female-defined ethnicity in Jewish populations?

One of the hardest things to get across to students in genetic history is that there is no right answer to a question of this sort. In fact, if a student working in my lab asks, "What should I do with the data?" the only honest answer I can give is, "Look for something interesting." It is an answer that students do not find completely satisfying, at least judging by the perplexed, your-hair-is-on-fire looks they give me. This is not to say that genetic history is nothing but creative interpretation. Genetic history that is researched accurately and intelligently depends on increasingly sophisticated tools developed in allied disciplines, most notably population genetics. The analyses borrowed from these disciplines must be carried out correctly, and they must be interpreted with caution and a sense of fairness. This is no easy trick, and anyone unable or unwilling to come to grips with the algebra and detailed conceptual structure of population genetics is well advised not to venture into genetic anthropology.

Because we don't know the history at the outset, however, we don't know which analyses will be instructive and which irrelevant. Hence, there is nothing to do but get to know the data and see what they have to say. This aspect of genetic history is in sharp contrast to some other areas of genetic research. For example, chapter 5 illustrates how medical genetics is now using many of the same population-genetic tools employed to study genetic history. The difference is that in medical genetics you often know what you are after. In my own work on the genetics of epilepsy or infectious diseases, for example, I study patients who do and do not respond well to a particular treatment or I study individuals who do and do not naturally control the virus responsible for HIV/AIDS. There is little ambiguity about what I am looking for in that context. I want to find a variant that will predict which medicine a patient should be given or a variant that might tell us what part of the immune system is most important in allowing some patients to adequately control HIV and never become sick despite being infected. But in the case of genetic history, scientists must have an open mind and assess whether the genetics can say anything at all of interest about the history. Genetic history is both art and science.

When it came to the study of Jewish Y-chromosome and mito-chondrial DNA variation, my colleagues and I thought we knew what needed to be done. In fact, it seemed quite straightforward. The idea was that female-defined ethnicity would block or retard gene flow from host populations into Jewish populations along the female, but not the male, line. We had typed Y-chromosome and mitochondrial DNA variation to reflect male and female patterns of movement, re-spectively. The obvious thing to do was to estimate the movements of Y chromosomes and mitochondrial DNA from the host populations into the Jewish populations. If our estimates of male and female ad-mixture rates, as they are called, were significantly different, we would have our evidence that female-defined ethnicity shaped the pattern of genetic variation in these populations. But if the estimates were the same, a long history of matrilineal descent might be nothing more than, to coin a phrase, an old wives' tale.

Admixture rates—the extent to which genes move from one popu-lation to another—can be estimated in various ways with varying degrees of sophistication. Neil and I looked at the results of a series of relatively simple estimates of admixture rates. To our disappointment, there were no obvious patterns and no striking differences between the Y-chromosome and the mitochondrial DNA data. Basically, the admixture estimates were consistent with a broad range of differ-ent patterns of movement and didn't tell us much about the role of female-defined ethnicity in Jewish history.

We went round and round as to what to do next. Should we use newer, more powerful statistical methods? Should we collect more data? Or should we look at the data we had in a different way? All we knew was that the summaries we had looked at thus far were directed toward the estimation of admixture rates and these had gotten us exactly nowhere.

One rainy weekend in London (are there any others?) we got together to look over another battery of simple summaries of the data that are routinely done in genetic history projects. After some discussion, it occurred to us that instead of focusing on admixture with host populations we should be looking at the peculiar pattern of mitochondrial DNA variation in the individual Jewish populations themselves. Some of the populations had mitochondrial DNA compositions so unlike those of the other populations we had studied that the methods used would certainly fail to determine the origins of the specific mitochondrial DNA types. Among the Georgian Jews, for example, we observed that more than half of all the individuals had a single mitochondrial DNA type, and that single type was hardly observed anywhere else in our data set (or in the rest of the world, for that matter), showing up only three times out of more than a thousand sequences. The Moroccan Jews showed a similar pattern, but with a different mitochondrial DNA type; the same was true for the Bene Israel.

I wish I could say it was a eureka moment, but it was actually more of a "duh" moment. We had been looking at estimated admixture rates, but looking directly at the underlying data made it blindingly obvious what was happening. Embarrassingly obvious. We then looked over some very simple and standard population genetic summaries of the data, and in about five minutes it was all over. The Jewish populations *always* had significantly less mitochondrial DNA diversity than the host populations and, in some cases, dramatically less.

Okay, you might say, so the Jewish populations had less genetic variation than the other populations. What does that have to do with

Jewish mothers? Aren't Jewish populations supposed to have under-gone sharp changes in population size and so on? That is the beauty of having both the mitochondrial DNA and the Y-chromosome data for comparison. Astoundingly, the Y chromosomes showed hardly any differences between the Jews and their hosts in terms of genetic diversity. This was not something that had affected populations as a whole; this was something that had *affected only the females.* We had finally figured it out: this story was not about admixture rates at all; rather, it was about a strikingly different demographic history for women and men in Jewish populations.

At some point in their histories, something important had hap-pened in the females of the nine populations we studied that had resulted in the loss of most of the mitochondrial DNA types found in the Eurasian host populations. And whatever happened, it happened more than once—in fact, it happened at least once in each of eight of the nine different populations. So we are talking about a minimum of eight events that reduced genetic variation for the female side of these populations and not the male side. We could tell this because when we looked at each of the populations in detail, we saw that they all had low mitochondrial DNA diversity compared with that of their hosts, but in different ways. Among the females of those nine Jewish populations were at least eight extreme founder events or population bottlenecks.[9]

A founder event is the founding of a new population by one or a small number of individuals. An extreme example, affecting the ma-ternal genetic heritage, would be if colonists in a new area took for themselves a number of sisters as wives; the offspring would all carry the mitochondrial DNA of the mother of those sisters. A population bottleneck occurs when a huge proportion of a population is prevented from reproducing (often by death). A classic example is the European bison (the wisent), whose population of thirty-six thousand is de-scended from a mere twelve individuals. Poaching in the eighteenth and nineteenth centuries created the population bottleneck that shrank the bison's gene pool. The net effect of founding events and bottlenecks is the same: a limited amount of genetic diversity.

All of this historical information could be seen not in sophisticated estimates of admixture rates but rather just by counting—tallying up the number of different mitochondrial types in the Jewish and host populations. Even in genetic history, sometimes "keep it simple, stupid" is good advice.

A skeptical reader might now wonder whether we can really pin this pattern on female-defined ethnicity or indeed any cultural features particular to Jewish populations. One way to address this question is to ask whether similar patterns are observed in other populations. I have tried to explain the difficulties in making population-genetic inferences through direct comparisons between mitochondrial DNA and Y chromosome variation—the apples-to-oranges conundrum. But one inference from such comparisons is becoming widely accepted. In 1998, Mark Seielstad, Eric Minch, and Luca Cavalli-Sforza compared the geographic pattern of Y-chromosome and mitochondrial DNA variation in global populations. They used a simple measure of genetic distance to assess the differences among these global populations for both genetic systems and compared it to geographic distance. They found that differences among populations in Y-chromosome composition build up very quickly with geographic distance, whereas mitochondrial DNA compositions remain much more similar over large geographic distances. In other words, even for populations located very near one another, the Y chromosomes are often sharply different from one another while the mitochondrial DNA is nearly identical.

As I discussed above, many differences exist between the Y-chromosome and mitochondrial DNA, and it is not easy to know which of the differences between the two might be responsible for the geographic pattern of variation reported by Seielstad and his colleagues. But the hypothesis they put forward is that the pattern results from patrilocality. In patrilocal societies the man stays put and the woman moves from her place of birth to join her partner in starting a family. Even if women move only small distances to start a family, say to neighboring villages, over time their relocating will have the effect of transferring mitochondrial DNA types over long distances. And if the men do not move, differences will build up on the Y chromosome—the diverse population of chromosomes will hang around and persist.[10]

At first, it seems an unlikely explanation. After all, isn't it men who settle new areas? From Roman legions to Viking raiders to European settlers in the New World, the image we have is one of men on the march or under sail. But this perspective is clearly biased by the historical period under consideration. It would appear that, for the bulk of our past, things have actually been quite different. In fact, it is estimated that in more than 70 percent of traditional societies today, it is the *woman* who moves when new families are begun. In most cases, the Y chromosome shows greater differences among populations than does the mitochondrial DNA, thereby supporting this idea.

But not so with Jewish populations. Here we see the exact opposite pattern. When Neil and I compared Y-chromosome and mitochondrial DNA variation, we found that the mitochondrial DNA showed much greater differences. Critically, we also made the same comparisons among the host populations from the same geographic areas as the Jewish ones—they show the usual pattern of greater Y-chromosome differences. Thus, the Jewish pattern does not depend on any particulars of the geographic areas the groups come from but rather has something to do with the groups themselves. Their culture has given them a distinct pattern of variation on the male and female lines of genetic ancestry—one that suggests it has been the men who've gone mobile.

Individual genotypes—that is, one person's DNA signature at any place or places in the genome—rarely carry much information about geographic ancestry. Why? Populations are generally similar and rarely show more than slight statistical differences. But the sharp differences in mitochondrial DNA composition among the nine various Jewish populations we studied meant that we suddenly had a window on these people's geographic ancestry. To see this, consider the frequencies of common mitochondrial DNA types in different populations (for the variable control region). In most Eurasian populations, the most frequent, or modal, haplotype is known as the Cambridge Reference Sequence (for Cambridge, England, where it was the first one to be completely sequenced). This sequence has a frequency ranging between about 5 and 15 percent in most populations. In our study, it ranged from about 4 to 12 percent in the host populations—nothing unusual there.

But in seven of nine Jewish populations, the modal type was not the Cambridge Reference Sequence but rather one rarely seen in non-Jewish Eurasian populations. Even more striking, the average frequency of the Jewish modal types was over 22 percent, while some of the Jewish groups had much higher frequencies than that. In the Georgian Jews, for example, over 50 percent carried the most common type. In the Bene Israel and the Moroccan Jews, the figures were 41 and 27 percent, respectively. And yet we see no such pattern in the Y chromosomes. The modal frequencies in the Y chromosomes in Jewish populations (which averaged 15.2 percent and ranged from 7.4 percent to 31 percent) were not significantly different from those seen in the host populations (mean 13.6 percent, range 8.1 percent to 31.2 percent). The fact that the most common mitochondrial DNA types were not what are typically seen in Eurasian populations means that in some cases we could make a reasonable guess about geographic ancestry based solely on an individual's mitochondrial DNA type. Thus, someone of Jewish ancestry carrying the Georgian Jewish modal mitochondrial DNA was very likely of Georgian Jewish extraction on the maternal side.

These differences also suggest that mitochondrial DNA variation could be used as a sensitive indicator of contact among Jewish populations. For example, migration of women between the Georgian Jewish community and the Bene Israel would very likely result in the transfer of the modal types between these groups.

How Did It Happen?

What our data say is that a sharp difference exists between the maternal histories of Jewish populations and their hosts. We can be fairly confident that there must have been severe bottlenecks or founder events in the female histories of some of these populations, but not in the male histories, because we see no reduction in Y-chromosome variation. And we can say even more than that. Because each different Jewish population has its own most common mitochondrial DNA type, these founder events must have occurred independently—at least one time in the history of eight of the nine populations. Recent work by Doron Behar, Karl Skorecki, and their colleagues has identified four founder events occurring two to three thousand years ago and accounting for some 40 percent of the mitochondrial DNA lineages seen in Ashkenazim today. Although the ultimate origin of their four founding mothers is unclear, the theme remains the same: Jewish women traveled—or did not travel, as the case may be—a demographic path through history quite different from men's.

What could be behind the dramatic differences in male and female genetic history? We will probably never know exactly. The problem is that, except for the Cambridge Reference Sequence, the types common in the Jewish populations are not common anywhere else. Thus, in any random sample of a hundred individuals (excluding Georgian Jews), the Georgian Jewish type is seen at most once or twice. But seeing something one time in a hundred is not statistically different from seeing it zero times in a hundred. So as far as we can tell, the Georgian Jewish mitochondrial DNA type could have come from anywhere in the world.

Still, we can speculate about a history that could explain what we see. The results are consistent with a story of Jewish men, perhaps traders along the Silk Road or the Arabian Peninsula, traveling long distances to establish small Jewish communities. They would settle in new lands and, if unmarried, take local women for wives. The communities might have been augmented by additional male travelers from Jewish source populations. Once they were established, however, the barriers would go up against further input of new mitochondrial DNA precisely because of female-defined ethnicity; few females would be permitted to join. So the genetics have given us a model for the demographic origin of Jewish communities, and it has some strange aspects to it. These communities began with Jewish males and, presumably, non-Jewish women, and not many at that. Once the Jewish community was in place, however, it disallowed further involvement of women from other communities, Jewish or otherwise.

Of course, we cannot say with certainty that this is what happened, but it would explain the genetic data. Moreover, it is consistent with narratives of isolation and endogamy associated with a number of Jewish and Judaizing groups, including the Bene Israel, the Bukharans, the Georgians, the Persians, and the Lemba.

The implications of founder events in the female histories of Jewish populations go beyond history—they may also have real-world consequences. For example, a company in Iceland called deCODE Genetics is trying to exploit the unusual population history of that island nation. Iceland was colonized by relatively few Norwegian settlers more than a thousand years ago. DeCODE and other scientists therefore expected the Icelandic population to be genetically homogeneous—remember, the consequence of the founder effect is less genetic diversity. If that was true in Iceland, it could help in efforts to identify mutations in genes that predispose individuals to particular diseases. This is because in a homogenous population the genetic underpinnings of complex genetic diseases should be easier to detect than in more genetically diverse populations. Why? The associations between genes and disease will point in many different directions in the heterogeneous populations but in fewer directions in the homogeneous ones. In other words, if finding a disease gene is akin to looking for a needle in a haystack, using a genetically homogeneous population means searching through a much smaller haystack to find the odd-looking needle.

But it is by no means clear that the founding event among the Icelandic population was sufficiently severe to substantially change the pattern of genetic variation. For example, if you compare Y-chromosome and mitochondrial DNA variation in Iceland with that in Europe, the differences are slight or nonexistent. This is not to say, however, that deCODE has been unsuccessful. While Icelanders do not appear to be all that much less diverse genetically than other populations, a striking proportion of new genetic discoveries are due to the deCODE team.

I chalk these discoveries up to the quality of the genetic work being done by deCODE and I suspect they'd make these discoveries based on other population samples, too, if they had the same team and critically the same clinical data. It does look, though, like there really could be some populations out there with more pronounced founder effects. Georgian Jews, for example, have mitochondrial DNA diversity that is about half of the average found in Europe. In the case of Iceland, the simple story of Viking settlers overlooked Irish women: genetic analyses suggest that the Vikings brought a lot of them to Iceland, thereby significantly augmenting the Icelandic gene pool and making it look much more European. It seems the haystack may not be that much smaller after all. Jewish populations, on the other hand, might be a particularly valuable group for genetic study.

CHAPTER 5

Look on Mine Affliction:
Genetic Diseases and Jewish History

Why is my pain perpetual, and my
wound incurable, which refuseth to be
healed? Wilt thou be altogether unto
me as a liar, and as waters that fail?

Jeremiah 15:18

The Jewish share of the overall burden of genetic disease
is probably no greater than that of Africans, Arabs, Icelanders, or anyone else, but I wouldn't blame you if you came
away from the biomedical literature with a different impression. A query of the PubMed database (www.pubmed.gov)
in early 2007 for the term *Ashkenazi* in titles and abstracts,
for example, retrieved more than thirteen hundred papers.

Contemporary Jews are primarily urban, they tend to
reside close to major clinical and research centers, and
they have a long history of involvement in the research
enterprise. Thus, the popular notion of "Jewish genetic

disease," as if one inherited certain genes only if one's family kept kosher or observed the Sabbath, is misleading: Jews' cultural and demographic tendencies have a lot to do with their coming to the attention of doctors and scientists—this is what we call an ascertainment bias.

The concept of Jewish genetic disease, however, is also rooted in some of the same historical and demographic forces that preserved the Cohen Modal Haplotype, sowed the peculiar Levitic Y-chromosome signature across Ashkenazi Europe, and gave rise to the various mitochondrial DNA patterns one finds in Jewish women down through the ages. As with all other genomes, a number of factors—reproductive behavior, population bottlenecks, and natural selection—have had a hand in shaping the anatomy of the "Jewish" genome.

As a genetic historian, I am interested in understanding where and when so-called Jewish disease mutations originated. Here I explore those questions, focusing on the implications, both medical and nonmedical, of those mutations for the people who carry them.

When we survey the world's 15 million Jews, can we say whether genetic diseases are more frequent among them? As with most of the other questions I've asked in this book, the answer is a qualified one. Some diseases are more frequent in some Jewish populations. At least forty conditions determined by single genes (Mendelian disorders) have been described in Jewish populations.[1] Perhaps half of those occur at higher frequency among Jews than among a random sample of the world's 6.6 billion people. Is there some reason for that observation?

There are dozens of disease-causing mutations found among Jewish populations. Defective genes predispose unlucky people to Tay-Sachs, Gaucher disease, breast and colon cancer, Parkinson's, Canavan disease, adrenal hyperplasia, Crohn's disease, dysautonomia, hypercholesterolemia, glycogen storage diseases, cystic fibrosis, and familial Mediterranean fever, among many others. Each one has its own clinical, genetic, and historical story. You can learn more about these conditions by consulting the references at the end of this chapter or checking out online Mendelian Inheritance in Man (http://www.ncbi.nlm.nih.gov/entrez/query.fcgi?db=OMIM), a massive catalogue of mutations that result in Mendelian diseases.

Why and how did all these diseases become so entrenched among Jewish populations? On several visits to Tel Aviv, I have decamped with my colleague Uri Seligsohn to the restaurant next to Tel Hashomer Hospital, where he works as a hematologist, to discuss this very question. I don't know that we've ever reached a definitive answer (or that we ever will), but I always enjoy the conversations, to say nothing of the massive Israeli salads. In any case, there are two prevailing theories, and both deserve thoughtful consideration. One of them could have implications not only for disease but also for some of the cultural and behavioral traits we may associate, rightly or wrongly, with Jewishness.

The debate comes down to this: drift versus selection. Genetic drift, as you'll recall, can result from a small population's settling in a new area, causing the eventual population to be genetically different from a larger source population and dependent on the precise genetic makeup of the small number of founders of the new group. Recall the example described in chapter 4 of a group of male colonists all marrying a group of sisters and the offspring marrying among themselves (this example, of course, is too extreme because of inbreeding, but the point is the same with slightly less related founders). The mother of these sisters will make an outsize contribution to the DNA of succeeding generations. As we know, her mitochondrial DNA will be the only mitochondrial DNA present. Some of the rest of her DNA will be diluted out by the process of evolution and the reshuffling of the deck with each conception, but much of it will remain.

The proponents of genetic drift argue that the high frequency of assorted disease mutations we see today among Jewish groups can often, if not always, be explained by the demographic peculiarities of both ancient and more recent Jewish populations. Of great importance is history. Clearly there is a long and bloody trail of Jews enduring wars, pogroms, and pandemics. From the Romans to the Nazis, Jewish populations have suffered devastating attacks that often resulted in sharp reductions in population size. These so-called bottlenecks, where much of the population is prevented from reproducing, can reduce the variety of genes that are passed on to future generations.

Other factors have contributed to a functional shrinkage of the Jewish gene pool. The exhortation to marry within the faith, first made so passionately by the Old Testament scribe Ezra, was codified in the Talmud and remains an underpinning of contemporary Judaism in many sects. Certainly plenty of American Jewish adolescents hear from their parents about the disastrous consequences of taking up with a Gentile. For hundreds of years this tradition of marrying within the faith, coupled with Judaism's relinquishment of proselytization in the early centuries of the Common Era, meant few if any new Jews coming to the faith were not conceived by other Jews. To this day, Judaism does not actively seek converts.[2] Finally, perhaps as an outgrowth of Judaism's endogamous traditions, consanguinity was, for a long time, a common marital practice among Jews. As recently as 1960, 1.4 percent of Israeli Ashkenazim married blood relatives (typically first or second cousins); among Iraqi Jews the consanguinity rate was nearly 30 percent. Traditionally, Jewish groups have not regarded consanguineous marriages as taboo.

This, in essence, is the case for drift: a shrunken gene pool has led to the fixation of certain mutations in Jewish populations. Fair enough. There appears to be strong evidence that multiple mutations in certain diseases, such as dysautonomia, came about due to founder effects.[3] But is there an alternative explanation for at least some mutations?

The alternative to drift is a Darwinian one. In high school biology, most of us were taught about the sickle-cell trait. Persons with two copies of a mutant beta globin gene have red blood cells that contort, or sickle, causing pain in the joints, infections, anemia, night sweats, cough, and fatigue. This is the autosomal recessive disorder sickle-cell anemia. By contrast, persons with only one defective copy of the gene are normal carriers of the sickle-cell trait. Not only are they normal, they are resistant to malaria—carriers of the mutations have blood that is inhospitable to the malaria parasite. Consequently, in areas of the world where malaria is prevalent, the mutation carriers have an advantage. In such environments, these types of mutations can be positively beneficial and therefore increase their representation in the population over time. In parts of Africa, the sickle-cell-carrier frequency may be as high as 50 percent. The same argument made for sickle-cell anemia—that carriers have an advantage that keeps the mutation (or mutations) in the population—has been made for a number of genetic diseases common among Jews.

The contrast between the two schools of thought is clearly reflected in discussions surrounding the mutation responsible for idiopathic torsion dystonia, a disease characterized by involuntary movements. The mutation responsible for this condition has been traced to a small part of Byelorussia and the pro-drift camp argues that pogroms and other demographic upheavals drastically increased the frequency of the mutation from its single origin about 350 years ago. The pro-selection camp sees the hand of selection, arguing that the condition is associated not only with involuntary movement but also with intelligence. Not a coincidence, they say.

Perhaps nowhere has the debate been more contentious than with regard to lysosomal storage diseases. Before wading into that argument, let us take a closer look at one of those diseases, Tay-Sachs. A child is born to a Jewish couple. For the first six months the baby is the picture of health, and then everything begins to go wrong. The child's muscle tone deteriorates—he or she becomes floppy and unable to support him- or herself. He or she stops responding to light and loses the ability to swallow. The situation worsens: the child experiences seizures, dementia. Eventually he or she is mentally retarded, paralyzed, and completely unresponsive. By age four the child is dead from pneumonia.

The first child of Josef Ekstein and his wife, a boy, was born with Tay-Sachs disease in 1965. Eighteen years later, Mrs. Ekstein gave birth to a baby with Tay-Sachs—*for the fourth time.* "I don't know how I got through it," Ekstein, an Orthodox rabbi in New York City, told *New Scientist* in 2004. "My strategy to overcome this tragedy was my faith, and to look forwards, not backwards." Today, thanks in large part to a screening program set up by Ekstein, Tay-Sachs is virtually nonexistent among the Orthodox Jewish communities in New York and Israel.

How do Tay-Sachs and other lysosomal storage diseases wreak their havoc? Lysosomes are the waste-processing centers of our cells. They are membrane-bounded sacs chock-full of enzymes whose job it is to break down bacterial remnants, used-up cellular parts, and assorted other detritus the cell can no longer use. When even a single one of those forty enzymes fails because the gene that encodes it is defective, the result is a disease such as Tay-Sachs. The molecules that the defective or missing enzyme was supposed to degrade instead accumulate inside the lysosome, causing all sorts of problems. In addition to Tay-Sachs, three other lysosomal storage diseases occur at higher frequencies in Ashkenazi populations.

Tay-Sachs disease is an autosomal recessive disease; that is, each parent of an affected child carries a single defective copy of the hexosaminidase A (HEXA) gene. In any given conception, each carrier parent has a 50–50 chance of passing on the defective copy to the child. In the event the child receives a defective copy from both parents (a 1-in-4 chance if both parents are carriers), then he or she will have Tay-Sachs disease. In Rabbi Ekstein's case, he and his family were the victims of bad luck: the probability of two carriers conceiving four (consecutive) babies with the disease is 1 in 256.

What makes Tay-Sachs a Jewish disease? It is found a hundred times more frequently in infants of Ashkenazi ancestry (or at least it was until widespread screening took hold). Among North American Jews, one child in thirty-one carries a Tay-Sachs mutation. In a common pattern among recessive diseases, a single mutation in the HEXA gene accounts for the vast majority of Tay-Sachs disease (73 percent). Two other mutations explain another 20 percent of the disease, with the remainder due to rare mutations (nearly eighty mutations have been described in all).

Interestingly, a distinct HEXA mutation is common among Moroccan Jews. That a mutation different from the one carried by Ashkenazim is prevalent among Moroccan Jews suggests that Tay-Sachs disease originated in these populations after the destruction of the Second Temple in 70 C.E. but before the rise of the Ashkenazim in 800–1000 C.E. Work reported in 2004 by geneticist Leah Peleg and her colleagues at Tel Hashomer Hospital in Israel (Karpati et al., "Specific Mutations in the HEXA Gene") identified a new mutation among Iraqi Jews and suggests that the disease we see today may actually have originated among post-exilic Jews (about 442 B.C.E.).

Given the horror of a disease like Tay-Sachs disease, population geneticists have wondered why it has persisted in the human gene pool. Are Tay-Sachs and other lysosomal storage diseases seen at higher frequencies among Jews—Gaucher disease, the most common single-gene disorder found in Ashkenazim; Niemann-Pick disease; and mucolipidosis type IV—simply the result of random drift or, like the sickle-cell trait, do they confer an advantage to those who are unaffected carriers? For Tay-Sachs itself some have argued that carriers may have increased resistance to tuberculosis, but the evidence has never been strong.

In fact, the debate pitting these two explanations against each other is not new. The question of lysosomal storage disease has been raging for forty years—since long before the advent of molecular genetics and the genome age. To those who favor selection, the inescapable fact is this: four distinct lysosomal disorders—Tay-Sachs, Gaucher, Niemann-Pick, and mucolipidosis type IV—occur at increased frequencies in the Ashkenazi population. Each one is part of the same biochemical pathway, and each disease's underlying gene is involved in the storage of specialized lipid molecules known as glycosphingolipids. For four related diseases to arise independently in the same population seems unlikely on its face. With his usual elegance, Jared Diamond in 1994 likened the mutations that cause these to lightning strikes. If lightning hits your house once and not your neighbor's, it's bad luck. If it happens repeatedly, however, you need to wonder why. Many have viewed the lysosomal storage system as too often hit by lightning to be explainable by chance.

Thus, those in the selection camp say, there must be a selective advantage for carriers of these mutations. Their argument is bolstered by the observation that multiple mutations can be found in each disease, leading, for example, to the repeated lightning strikes of Jared Diamond's 1994 article. If the cause were founder effect, they say, wouldn't we see only one primary mutation? We don't.[4] Moreover, selection among Jewish populations is not without precedent. Like sickle-cell carriers, carriers of certain other mutations affecting the same protein are associated with beta-thalassemia, a severe anemia, and are also more resistant to malaria. Kurdish Jews, for example, exhibit a high frequency of thalassemia (Kurdistan is a malaria-prone region); the trait has almost certainly been selected for in the Kurdish population.

The pro-drift forces counter that lysosomal storage diseases are hardly the only class of diseases seen at high frequency in Jews. What about cancer? Three tumor-suppressor genes, BRCA1, BRCA2, and APC, all cause inherited cancers at relatively higher rates in Jews, with certain mutations in these genes more common in Jews than in other groups. Is there some advantage to having these mutations, too?

What about the clotting disorders? The "drifters," led by Neil Risch, a distinguished human geneticist at the University of California in San Francisco, argue that what we're seeing is a detection or ascertainment bias: they believe the recognition of a lysosomal storage disease or two among Ashkenazim called attention to patients with related diseases. They also contend that the true number of mutations in lysosomal storage diseases is not really different from the number of mutations seen in other types of genetic diseases. They go on to wonder why Ashkenazim are unique. As Risch and Hua Tang and their colleagues wrote in 2003: "It seems implausible that all these disease mutations have undergone selective advantage unique to one population."

In reality, neither side is arguing for an all-or-none phenomenon. Israeli geneticists Joel Zlotogora and Gideon Bach, at the leading edge of the pro-selection camp, are not saying that selection alone is responsible for what we see today. They acknowledge that small founder populations were a critical factor in increasing the frequency of mutations seen in Jewish diseases. Likewise, some drifters do not rule out selection; they are simply waiting for data that prove that a particular environmental pressure selected for the high prevalence of any particular mutation among Ashkenazim. Of course, they may be waiting a long time as positive selection is a notoriously hard thing to prove.

The reader may say, Yes, yes, this is all well and good, but isn't the selection-versus-drift-in-Ashkenazim debate really just a minor tiff of interest only to a few pointy-headed geneticists? Until recently, I might have agreed. In 2005, however, the media picked up on a paper by anthropologists Gregory Cochran, Jason Hardy, and Henry Harpending that has effectively moved the argument from the realm of science to that of sociology. Now we geneticists have a genuine kerfuffle on our hands.

What Cochran, Hardy, and Harpending proposed is that the unique demographic and sociological history of Ashkenazim in medieval Europe created an environment that made intelligence a highly advantageous trait. As part of this process, they hypothesized, certain gene variants that conferred intelligence were selected for in Ashkenazim. In particular, carriers for mutations in lysosomal storage genes were likely to exhibit increased intelligence. In other words, not only were lysosomal storage diseases selected for but they gave Ashkenazi Jews an intellectual advantage.

Just as two copies of a mutant globin gene will cause sickle-cell anemia, Cochran, Hardy, and Harpending noted, two mutant copies of a lysosomal enzyme gene will cause Tay-Sachs disease, Gaucher, Niemann-Pick disease, and mucolipidosis type IV. No one argues that point. But the authors went further: just as one copy of the same globin gene will protect the carrier against malaria, they contended, so too does one copy of the lysosomal enzyme gene enhance intelligence.

As one can imagine, the paper caused quite a stir. To support their claims, Cochran, Hardy, and Harpending pointed to the fate of Ashkenazim in medieval Europe. Banned from guilds and forbidden to own land, Jews were forced to be resourceful in order to make a living. Because Christians were prohibited from engaging in usury, a large proportion of Jews wound up working as moneylenders. Without question, the job required intelligence and mathematical literacy. Could it be that the main occupation Jews were consigned to acted as a selective force favoring intelligence?

In addition to presumptive selection and the idea that four lysosomal storage diseases prevalent among Ashkenazim cannot be coincidental, the other leg of the Cochran-Hardy-Harpending argument is a contemporary one: a relatively large number of Jews are prize-winners and chess champs, score very well on standardized tests, and so on. Referring to the preponderance of lysosomal storage diseases among Ashkenazim and to the large number of Ashkenazi Nobel laureates, Cochran wondered aloud to *New York* magazine why 27 percent of Nobel Prize winners have been Jewish.

What about the biological evidence? Cochran, Hardy, and Harpending cited numerous studies of brains from individuals with lysosomal storage diseases showing anatomical differences, involving neuronal dendrites in particular, that might be suggestive of more interconnections among neurons.[5] The implication? Patients with lysosomal storage diseases have diseased but potentially more highly developed brains. The director of a Gaucher clinic in Jerusalem reported that nearly one-third of his patients held jobs that, according to Cochran, Hardy, and Harpending, required an IQ of 120 or better.

Their paper, published in a journal not frequently read by the human genetics community, is tantalizing, circumstantial, politically incorrect in the extreme, and entirely speculative. In my view, the possibility that glycosphingolipids have something to do with intelligence cannot be ruled out. But that's a very long way from being able to confidently invoke the carrier status of lysosomal storage disease mutations as an explanation for Ashkenazi intellect, if that's even what we're measuring. Consequently, there's nothing to say that it's the lysosomal storage disease genes per se that are conferring a selective advantage of one kind or another to carriers; if the advantage is due to genes at all, they may very well be genes located in the same chromosomal neighborhood as the lysosomal storage disease genes. And if so, those genes may have nothing to do with intelligence or any other cognitive function. One also has to wonder about the lysosomal storage disease mutations that have been found in Moroccan and Iraqi Jews who are not descended from moneylenders. Can we assume that the Sephardi carriers of those mutations are less intelligent than Ashkenazi carriers of mutations in the same genes? Were the Sephardim subjected to earlier or different selective pressures?

My other major qualm is about how to define intelligence when making these types of claims. Genetic study is impossible to do well without a clear understanding of what a trait is and what it isn't. This is the main reason that finding genes that predispose to psychiatric disorders has been so difficult. What is schizophrenia? What is depression? If we lump together conditions that are biologically different, it is that much harder to find the genes that influence them. So, what are Cochran, Hardy, and Harpending referring to when they talk about intelligence? Is it simply smarts as measured by IQ tests? Cleverness? Street smarts? Might it include less admirable traits—say, an amalgam of anti-Semitic stereotypes such as frugality, aggressiveness, and cunning like Shylock's? Such stereotyping and genetic determinism, in my view, permeated the controversial (and to me unpersuasive) 1994 book *The Bell Curve* and remain daunting occupational hazards for behavioral geneticists.

Even if we restrict ourselves to presumptively objective IQ-based definitions, intelligence, like behavior, remains a highly complex trait and there are plenty of data demonstrating that environmental factors play an essential role in determining it. Jews have a tradition of scholarship that is as old as the Torah. Education has always been a hallmark of Jewish culture. Are we prepared to attribute much or most of that scholarly impulse to genes? Is there a danger that the Cochran-Hardy-Harpending approach will somehow legitimize the pseudoscientific genetic determinism that we as scientists and citizens have spent so long fighting against? Genes are not destiny. Nor would I want to minimize the value of good old-fashioned hard work. As far as I'm concerned, our environment, our choices, our agency are factors we ignore at our peril. I hope that readers of this book will come away from it understanding that genes play a critical role in who we are but that they always do so in combination with the environment. And in human populations, genetic and environmental contributions are notoriously hard to disentangle. I do not dispute Cochran, Hardy, and Harpending's observation that Jews do particularly well on some sorts of cognitive exams and that in some societies they are hugely overrepresented in certain brainy disciplines. I am simply stating: consider the context. In addition to any small genetic differences, American Jews, for example, have had different familial, cultural, and environmental experiences from those of non-Jewish Americans. Forced to choose between Jewish mothers and Jewish genes, I'll take Mom every time.

Having said all that, I would point out an attractive feature of the Cochran-Hardy-Harpending hypothesis—it is easily testable. Their idea is that carriers of specific mutations are, on average, smarter based on standardized exams. For anyone really interested in settling this question, the appropriate approach is the obvious one: administer standardized cognitive tests to a large number of individuals who do and do not carry the relevant mutations. Although I would not put much money on the hypothesis being correct, who really knows? If it is, the implications are dramatic, and not primarily because of questions of Jewish history. Identifying gene variants that influence cognitive abilities could point the way toward new therapies for any number of neurological and cognitive disorders.

Despite the relative rarity of diseases commonly found in Jewish populations, genetic screening and diagnostics for these diseases have nevertheless become big business. Sometimes a population's genetic legacy can have profound financial consequences. Never has this been truer than it is for Ashkenazim at risk for familial forms of breast cancer.

In the mid-1990s, several academic and corporate labs engaged in a furious race to clone the BRCA1 and BRCA2 genes that predispose to familial breast cancer.[6] Myriad Genetics, a genetic diagnostics and pharmaceutical company based in Salt Lake City, won the race. The firm filed a series of patents based on the sequence of the genes. Much to the consternation of breast cancer researchers and clinical genetics labs everywhere, Myriad obtained a monopoly on BRCA testing: all clinical tests would have to be performed at the company's Utah labs.

To this day, Myriad (and the University of Utah) retain a virtual monopoly on BRCA testing in the United States. In Europe, however, legal challenges have whittled away at Myriad's patents. In mid-2005, the company's BRCA2 patent—now held by a consortium including the University of Utah—was restricted to 6174delT, the most common BRCA2 mutation and the one that appears almost exclusively in Ashkenazim. This means that European patients are now asked if they are of Ashkenazi extraction. If they answer yes, they are required to pay the same royalties (about five hundred dollars in 2005) for BRCA2 testing that patients in the United States pay.

What is the significance of this decision by the European Patent Office? First, it was terra incognita—European patent holders rarely demand royalties or license fees from public health laboratories and clinics. Outraged at the decision, a sizeable number of clinics opted to withhold fees. A coalition of European medical geneticists and governmental organizations has challenged the patent (and other Myriad patents) in court. Second, given the relatively high frequency of the mutation in Ashkenazim and the high cost differential between the EU and Israel (in Israel, the test was about seventy dollars in 2006), Israeli oncologists and geneticists have decried the ruling as racist.

I find it hard to disagree with that assessment. To charge seven times the cost for a medical procedure in one ethnic group versus another, whatever the arcane patent-law explanations, violates fundamental tenets of equality valued by any democratic and egalitarian society. Such policies could even risk exacerbating health care disparities among some ethnic communities (although, admittedly, Jews in Western countries are rarely disadvantaged in this regard). Myriad should not be deprived of a return on its investment in the BRCA2 gene, but technical requirements and the details of intellectual property should not be allowed to foster a society in which individuals are penalized for their genetic heritage. My hope is that this situation can be resolved in a just and equitable way, one that will set a global precedent as to where intellectual property ends and the rights of individuals, whatever their unique genetic footprints, begin. Our genetic heritage is ours to treasure, to explore and to marvel at, but we should never tolerate its being used against any person or any people.

CHAPTER 6

Jews, Genes, and the Future

And I will bring you out from the
people, and will gather you out of the
countries wherein ye are scattered, with
a mighty hand, and with a stretched
out arm, and with fury poured out.

Ezekiel 20:34

I am a geneticist. But I am also a Jew. As I intimated
in the preface, I would not suggest that it is purely coincidental that I wound up studying the genetics of
Jewish populations. Traveling to Israel to meet collaborators, reading about three millennia of history, and seeing landmarks that influenced, and were influenced by,
countless momentous events in human existence have
connected me to both human history and my own history.

It bears repeating yet again: genetic history is still
just history. In fact, it offers only the briefest and most

superficial of glimpses into history. But those glimpses are sometimes into pasts that are themselves deep and remarkable, and in my biased view none is deeper and more remarkable than that of the Jews.

The first recorded mention of the nation of Israel, on the stela commissioned by the pharaoh Merneptah in 1207 B.C.E., boasted of its conquest by the Egyptians: "Israel is stripped bare, wholly lacking seed." Poetic, yes, but perhaps a little premature. In no small way, it is the thirty-two hundred years' worth of narratives written and told since the pharaoh's presumptuous proclamation that has driven me to examine Jewish genetic history. My projects have encompassed modern-day priests of higher and lower castes, an African tribe claiming descent from the patriarchs, peripatetic Jewish women and men from every continent, Slavs and Turks, Ashkenazim, Sephardim, and Mizrahim (Jews from the East), and Orthodox and secular. I am both gratified and pleasantly surprised by the ways that modern genetics has been able to shed even a little bit of light on their stories.

For me and perhaps for the reader, the natural question is, What next? Well, one answer is, More of the same. There are new stories to be uncovered and told in genetic history. Beginning around 2005 a resource has become available that I expect to yield a quantum leap in our understanding of human genetic history, to say nothing of human health: the HapMap (http://www.hapmap.org). The HapMap represents a global assessment of haplotypes—sets of closely linked genetic variants on a chromosome that tend to be inherited together—in four human populations: persons from Utah descended from northern and western Europe, Japanese, Yorubans from Nigeria, and Han Chinese. The idea behind the HapMap is, in a small number of populations, to capture the lion's share of human genetic diversity. By typing more than one million variants (with millions more to come) in 269 people representing the four populations, the HapMap Consortium has created a tool kit that will give genetic historians a baseline measure of genetic variation. From now on, DNA from every population under study, not to mention every disease mutation, can be compared with HapMap data to gain a clearer sense of where genes predisposing to diseases and other traits have been, where they are at present, and where they might be going. I expect the HapMap, as well as millions of polymorphic markers that weren't available when I undertook this work a decade ago, to dramatically increase our resolution in viewing the genetic history of populations, just like a microscope outfitted with a more powerful lens. At the time of this writing, my lab has just gotten a very strong intimation of how much more resolution is becoming possible. Many groups are using these gene chips (as the probe sets allowing simultaneous typing of many polymorphisms are called) to study the genetics of disease and other human traits. These chips are able to genotype huge numbers of HapMap polymorphisms at the same time so that one gets information on hundreds of thousands of polymorphisms for each sample (and, indeed, indirect information about millions of other polymorphisms). Looking at this huge mass of genetic data as part of a study of cognitive and psychiatric genetics, my colleagues and I were astounded to see that it was possible to predict accurately those individuals claiming Jewish ancestry on the basis of their genetic composition alone. This work must be considered provisional as it has not yet gone through the rigors of peer review and publication, but it does suggest that gene chips will usher in a new era in genetic anthropology.

These new tools will help to uncover new stories in Jewish genetic history. As I touched on in chapter 2, the history of Ashkenazi Jewry remains the biggest of black boxes. Ashkenazim are the dominant Jewish population in the world today; five hundred years ago they were on the brink of extinction. Historian Paul Wexler maintains that they are fundamentally a Slavic and not a Germanic people. Arthur Koestler posited that they are Turkics descended from the mysterious Judaizing kingdom of the Khazars that stretched between the Black and Caspian seas a thousand years ago. Mainstream historians contend that they arose in and around Germany and France between the sixth and ninth centuries C.E. Well, which is it? Perhaps none of the above. Jews have lived in Rome continuously since the Hasmonean dynasty in the second century B.C.E. Modern Roman Jews claim to be neither Sephardim nor Ashkenazim. They view themselves (correctly, in my opinion) as truly "old school." To date, genetic studies have been unable to resolve the issue of their origins. More extensive haplotyping of the Romans and of classic Ashkenazi, Sephardi, and Mizrachi populations might clarify matters. By the same token, genetic studies of contemporary Turkic and Slavic populations might finally settle the nagging question of the Khazars' place in Ashkenazi history.

Although the Ashkenazim seem to get most of the attention, I am equally intrigued by the mysteries surrounding small, remote Jewish populations, many of which are described in Ken Blady's *Jewish Communities in Exotic Places*. The Djerbans, for example, have lived on an island off the coast of Tunisia perhaps for as long as twenty-five hundred years, practicing a pious brand of Judaism that has remained virtually unchanged. Djerbans resemble their Muslim neighbors and speak an ancient brand of Judeo-Arabic. Are they the descendants of the Jews from the Babylonian exile? Are they more closely related to the Berber tribes from nearby Morocco? Similarly, although analyses of Y chromosomes of Samaritans, a presumably ancient Jewish offshoot, support a Near Eastern origin, much more could be lurking in their genomes.

And what of the Mountain Jews? A group of Jews in Kurdistan may have maintained an agricultural presence there since the Babylonian exile. Culturally distinct from their Iraqi neighbors, these Jews continue to speak a neo-Aramaic tongue not so far removed from the language of the Talmudic sages. Meanwhile, Mountain Jews of a different sort have long lived—and continue to live—in Daghestan in the far eastern reaches of the Caucuses, perhaps as remote a locale as any Jewish community has known. In the nineteenth century they

encountered Jews in the Pale of Settlement and their religious practices consequently took on some Ashkenazi influences. What might
DNA analysis tell us about the ancient origins and evolution of this
isolated, combative tribe whose members speak a Persian-Hebrew-
Turkic language?

Beyond these more specific questions, the large-scale genetic analyses that are now possible may finally allow us to address quantitatively
just how separate Jewish populations have been from their host populations. So far we have had only the narrowest glimpse into the overall
similarity among the genomes of individuals who may or may not
consider themselves Jewish. But there are medical studies going on
all over the world in which hundreds of thousands of genetic markers
are being typed in thousands of individuals, and in some cases those
markers have been statistically associated with ethnic, racial, or even
religious affiliations. How will the Jewish populations compare with
other groups in these studies? We already know that the Y chromosomes of Jews living in Europe are more typical of Semitic populations than, say, German or Scandinavian ones. To what extent will
this be true for the rest of the genome? Will Jewish populations be
generally different from the hosts, even if the differences are slight?
What, if anything, might these differences mean? In the coming years
we shall see.

We have pried open a totally new window on who we are and where
we came from, but we are only just learning how to see through it.
Soon many people will know a great deal about their precise genetic
makeup and at least some of their decisions about lifestyle and opti-
mal medical treatment will be influenced, for better or worse, by that
knowledge. We will also learn a great deal more about our evolutionary history, both recent, as I have discussed here, and less recent.
Genetics and genomics are entering a new age of discovery.

Chapter 1. Keeping God's House

1. Skorecki, an observant Ashkenazi Jew who believes himself to be a Cohen, had an epiphany in synagogue one morning when he saw a Sephardi Cohen ascend the bimah to read the Torah. Could he share a common ancestor with this man? Skorecki contacted geneticist Mike Hammer at the University of Arizona. Eventually, they joined forces with Neil Bradman and, later, me.

2. It was actually the ninth of Av by the rabbinic chronology. To this day, Tisha B'Av (the ninth of Av) is observed by Jews as a day of mourning.

3. Whether the book of Chronicles genealogy can be trusted is not entirely clear; as Blenkinsopp observes in *Sage, Priest, Prophet,* in all likelihood it was a transparent attempt to legitimate Aaronite claims to the office of high priest.

4. As I discuss in subsequent chapters, the Ashkenazim are believed to have originated in various parts of Europe. The term *Sephardim* originally referred to Jews from Spain and Portugal, especially those descended from the Jews who were expelled from those countries in the late fifteenth century. In the modern Israeli vernacular, the word *Sephardi* is often used inaccurately to describe all immigrant Jewish communities from countries bordering the Mediterranean and throughout the Near East and much of South America. In any case, the two communities are known to have remained largely separate since the expulsion.

5. Microsatellites are also known as short-tandem repeat polymorphisms and are probably familiar to most readers as the markers used in forensics to identify the perpetrator of a crime by "matching" crime-scene DNA to that of a suspect or that present in a DNA database.

Chapter 2. Lost Tribe No More?

1. For two reasons we do not have access to so-called ancient DNA from the Israelites who lived two thousand plus years ago: despite what Michael Crichton describes in *Jurassic Park*, ancient DNA is very hard to recover from human remains (let alone from those of dinosaurs), and in Judaism there is a strong prohibition against disturbing such remains.

2. Some scholars doubt the existence of an important Hebraic kingdom at this date, led by David or anyone else. Although I am far from an expert in this area, my own view is that a disparate body of evidence does support the traditional scholarly view of the establishment of a United Monarchy of the Hebrews, or Israelite Kingdom, around this time.

3. As Hillel Halkin points out in *Across the Sabbath River*, while tradition usually refers to the "ten lost tribes," only nine tribes belonged to the Northern Kingdom of Israel and could have been exiled or otherwise resettled by the Assyrians. The northernmost of the southern tribes, the tribe of Simeon, was presumably absorbed into the dominant southern tribe of Judah.

4. Until the end of apartheid, black South Africans were often called Bantus, a term that is now regarded as offensive within South Africa but that continues to be used elsewhere.

5. Ethiopian Jews, originally called Falashas ("outsiders" or "strangers"), consider that term pejorative and prefer to call themselves "Beta Israel" ("House of Israel"). Between 1984 and 1999, some seventy thousand Ethiopian Jews emigrated to Israel, many during a series of state-sponsored airlifts. Twenty thousand remain in Ethiopia and continue to agitate for the right to move to Israel. Many Ethiopian Jews regard themselves as descendants of King Solomon and the Queen of Sheba. A more detailed consideration of their story is taken up in chapter 4.

6. Crypto-Jews were once called Marranos ("swine" in Spanish). They were Jews from Spain and Portugal who were forced to convert to Catholicism in the fourteenth and fifteenth centuries. Many took refuge in Muslim and Protestant countries. Others continued to practice Judaic rituals in secret.

Chapter 3. Looking Out for Number Two

1. A presumptive descendant of Eleazar (third son of Aaron), Zadok was a high priest (a Cohen) in the time of David and Solomon. Both Zadok and his contemporary Abiathar served as priests to David (1 Samuel 22; 2 Samuel 8:17, 20:25). He was eventually dismissed from office for participating in an effort to enthrone David's eldest son, Adonijah, rather than Solomon (1 Kings 2:26–27). The Levites may have derived from a family of which Abiathar was a member. Meanwhile, the high priesthood remained in Zadokite hands for some eight hundred years until the rise of the Maccabees.

2. Genesis 34 recounts the story of Levi and his brother Simeon slaughtering every male resident of the village inhabited by Shechem, who raped Levi and Simeon's sister Dinah. Years later, on his deathbed, Jacob recalled the slaughter and cursed Simeon and Levi's anger, prophesying that his murderous sons' descendants would be dispersed throughout all the tribes of Israel (Genesis 49:5–7). In the eyes of many believers, so they were.

3. Deuteronomy makes no such distinction: there, all priests are referred to as Levites.

4. This ritual, *pidyon ha-ben* ("redemption of the son"), is described in Numbers 18:15–16 and is still practiced by many observant Jews who "redeem" their firstborn sons after one month by paying a Cohen five silver dollars. Such a deal.

5. According to a genealogy in Genesis 10, Ashkenaz was also the son of Gomer and grandson of Japheth.

6. The Pale of Settlement, established by Catherine the Great in 1791, was a western border region of Imperial Russia where Jews were permitted to reside. At its largest, it included much of present-day Lithuania, Belarus, Poland, and the Ukraine, as well as parts of western Russia. Pogroms, double taxation, and other depredations led millions of Jews to emigrate from the region to the United States and elsewhere. In the wake of the Russian Revolution in 1917, the Pale was finally abolished.

7. Without knowing the historical population sizes of the Levites it is difficult to determine how much genetic drift might have occurred.

8. Ironically, the Khazar theory has been seized upon by anti-Zionists as evidence that the Jews have no historical right to the land of Israel.

9. Jean-Baptiste Lamarck (1744–1829), the French naturalist, put forth the idea that physical traits acquired during one's lifetime could be passed

on to the next generation. Although Darwin's work all but discredited Lamarck's, Darwin praised Lamarck for calling attention to ideas of organic evolution.

Chapter 4. Those Jewish Mothers

1. Samaritanism is a very early offshoot of Judaism; indeed, some claim it is not an offshoot at all but rather the most continuous representation of early Hebraic worship. The Samaritans claim to be descendants of the tribes of Menasseh and Ephraim (sons of Joseph) and Levitical priests who were left behind during the Babylonian exile of 722–520 B.C.E. In the fifth century C.E., the Samaritans are thought to have numbered 1.2 million. Today, some 640 individuals remain in the Tel Aviv suburb of Holon and in the West Bank town of Nablus. A 2004 genetic study of living Samaritans by Marc Feldman, Peter Oefner, and colleagues (Shen et al., "Reconstruction of Patrilineages and Matrilineages") found evidence of ancient Samaritan and Cohen ancestry.

2. The Mishnah is the first written record of Jewish oral law. The Mishnah and the Gemara, a book of commentary on the Mishnah, together constitute the Talmud, although Gemara and Talmud are usually used interchangeably.

3. This is a point of some contention in the literature because not all geneticists make the same assumptions about mutation rates.

4. In population genetics, effective population size is the size of an ideal population in which one would see the same rate of evolution via chance (genetic drift) as one would in the real, natural population under study. In an ideal population, each individual has the same probability of contributing to the subsequent generation. Effective population size is typically smaller than census population size.

5. Iranian Jews were an exception. We did not have contemporary non-Jewish Iranian samples available for comparison.

6. In the event these populations trace more to the Arabian expansion associated with the spread of Islam in the seventh century C.E. and after, they could be thought of as representative of the nearby Arabian Peninsula.

7. Much of the material that follows comes from Ken Blady and S. Kaplan's excellent book, *Jewish Communities in Exotic Places.*

8. The paper was not peer-reviewed and should therefore be read with skepticism.

9. Founder events are a common feature of human populations and are thought to be important in the histories of, for example, the Icelandic and Finnish populations, the Parsi of India, and, more generally, non-African populations.

10. A 2004 paper from Mike Hammer's group at the University of Arizona (Wilder et al., "Global Patterns") found that, on a global or continental scale, the pattern described by Seielstad et al. breaks down. It may be that over short distances women tend to move around more but that when men move they tend to go a long way.

Chapter 5. Look on Mine Affliction

1. Transmission of these diseases follows the laws of genetics established by the nineteenth-century monk Gregor Mendel, the father of modern genetics.

2. Some would argue that the Orthodox and Conservative branches of Judaism, with their demand for elaborate rituals and onerous scholarship for those wishing to become Jewish, actively discourage converts.

3. Familial dysautonomia is a disorder affecting the development and survival of neurons in the autonomic nervous system. Symptoms include insensitivity to pain, an inability to produce tears, poor growth, and labile blood pressure.

4. As I've noted, though, many diseases common to Ashkenazim feature one mutation that accounts for a majority of the disease burden.

5. Dendrites, from the Greek word for tree, are branched filaments in neurons. It is the dendrites that receive most of the signals from other neurons via connections called synapses.

6. Today, in the postgenomic era, cloning a gene elicits a yawn—it is a prosaic and tedious act, especially since the entire human genome lives in a database accessible to anyone with a computer and an Internet connection. But in 1995 it was still a big deal.

allele. A specific form of a gene.

autosomal recessive. Any trait coded for by a single gene on a non–sex chromosome requiring two copies, one from each parent. Cystic fibrosis, for example, is caused by two faulty copies of the CFTR gene, one inherited from each parent.

base pair. Two nucleotides (DNA or RNA subunits) on opposite complementary DNA or RNA strands that are connected via hydrogen bonds.

chromosomes. Rodlike bodies resident in the nucleus of cells containing the chemical chromatin and a subset of an organism's genome (some organisms have only one chromosome). Their numbers are relatively constant in the cells of any one kind of organism. Humans have twenty-three pairs of chromosomes.

Cohen Modal Haplotype (CMH). A set of genetic markers that are usually inherited together on the male Y chromosome and that have been statistically associated with priestly origins among self-identified Jews.

DNA. Deoxyribonucleic acid, a nucleic acid found in cell nuclei and especially genes that is associated with the transmission of genetic information.

endogamous group. A community in which members generally mate within the group.

forensic DNA. One or more DNA samples obtained from a crime scene or mass disaster.

gene. A discrete stretch of DNA that can be transcribed into RNA that is usually (but not always) translated into protein.

genealogy. A historical account of the descent or ancestry of a person, family, or group.

genetics. The study of inheritance, often focused on just one gene or a few genes and/or traits.

genome. The entire DNA content of an organism. For example, the human genome consists of roughly three billion base pairs of DNA. The roundworm genome is roughly 100 million base pairs.

genomics. The study of genomes or parts of genomes, generally undertaken on a larger scale than traditional genetics.

haplogroup. A large set of haplotypes.

haplotype. A group of alleles within one or more genes occurring on a single chromosome that are close enough to one another to usually be inherited as a unit.

microsatellite. A polymorphic DNA marker that usually consists of repeating units of one to four base pairs in length. Because microsatellites exist in so many forms and can therefore be traced through multiple generations and kinships, they are especially useful for population genetic studies.

mutation. A permanent physical change in the sequence of DNA or RNA.

polymorphic markers. Genetic variants that exist in two or more forms. Polymorphic markers are essential for understanding how traits and diseases are inherited; they serve as signposts for those traits.

polymorphism. A genetic variant that exists in two or more forms. Polymorphisms are invaluable for genetic studies because of their association with particular traits.

regulation of gene expression. Control of the extent to which a gene is turned on or off. Gene expression is often regulated by DNA sequences that are not genes themselves.

RNA. A close chemical relative of DNA but used in a different way. Whereas DNA is primarily responsible for transmitting information between the generations, RNA acts as a message system within the body carrying instructions from the DNA about how to build proteins. Recently it has also been shown that RNA can have a variety of important functions in its own right.

Introduction

Carmeli, D. B. "Prevalence of Jews as Subjects in Genetic Research: Figures, Explanation, and Potential Implications." *American Journal of Medical Genetics, Part A* 130 (2004): 76–83.

Cavalli-Sforza, L. L. "The Human Genome Diversity Project: Past, Present, and Future." *Nature Reviews Genetics* 6 (2005): 333–340.

Cochran, G., J. Hardy, and H. Harpending. "Natural History of Ashkenazi Intelligence." *Journal of Biosocial Science* 38 (2006): 659–693.

Marks, J. "New Information, Enduring Questions: Race, Genetics, and Medicine in the 21st Century." *Genewatch* 18 (2005): 11–16.

Meadows, M. "FDA Approves Heart Drug for Black Patients." *FDA Consumer* 39 (2005): 8–9.

Redon, R., S. Ishikawa, K. R. Fitch, L. Feuk, G. H. Perry, T. D. Andrews, H. Fiegler, et al. "Global Variation in Copy Number in the Human Genome." *Nature* 444 (2006): 444–454.

Chapter 1. Keeping God's House

Blenkinsopp, J. *Sage, Priest, Prophet: Religious and Intellectual Leadership in Ancient Israel.* Louisville, Ky.: John Knox Press, 1995.

Bright, J. *A History of Israel.* 4th ed. Louisville, Ky.: Westminster John Knox Press, 2000.

Browne, L. *Stranger than Fiction: A Short History of the Jews from Earliest Times to the Present Day.* New York: Macmillan, 1925.

Dimont, M. I. *Jews, God, and History.* New York: Simon and Schuster, 1962.

Konner, M. *Unsettled: An Anthropology of the Jews.* New York: Viking Compass, 2003.

Rooke, D. W. *Zadok's Heirs: The Role and Development of the High Priesthood in Ancient Israel.* Oxford: Oxford University Press, 2000.

Shanks, H. *Ancient Israel: From Abraham to the Roman Destruction of the Temple.* Rev. ed. Washington, D.C.: Biblical Archaeology Society, 1999.

Skorecki, K., S. Selig, S. Blazer, R. Bradman, N. Bradman, P. J. Waburton, M. Ismajlowicz, and M. F. Hammer. "Y Chromosomes of Jewish Priests." *Nature* 385 (1997): 32.

Thomas, M. G., K. Skorecki, H. Ben Ami, T. Parfitt, N. Bradman, and D. B. Goldstein. "Origins of Old Testament Priests." *Nature* 394 (1998): 138–140.

VanderKam, J. C. *From Joshua to Caiaphas: High Priests after the Exile.* Minneapolis: Fortress Press, 2004.

Chapter 2. Lost Tribe No More?

Blady, K., and S. Kaplan. *Jewish Communities in Exotic Places.* Northvale, N.J.: Jason Aronson, 2000.

Chigwedere, A. S. *The Roots of the Bantu.* Marondera, Zimbabwe: Mutapa Publishing House, 1998.

Davis, D. S. "Genetic Research and Communal Narratives." *Hastings Center Report* 34 (2004): 40–49.

Halkin, H. *Across the Sabbath River: In Search of a Lost Tribe of Israel.* Boston: Houghton Mifflin, 2002.

Hammer, M. F., A. J. Redd, E. T. Wood, M. R. Bonner, H. Jarjanazi, T. Karafet, S. Santachiara-Benerecetti, et al. "Jewish and Middle Eastern Non-Jewish Populations Share a Common Pool of Y-Chromosome Biallelic Haplotypes." *Proceedings of the National Academy of Sciences of the United States of America* 97 (2000): 6769–6774.

Johnston, J. "Case Study: The Lemba." *Developing World Bioethics* 3 (2003): 109–111.

Le Roux, M. *The Lemba: A Lost Tribe of Israel in Southern Africa.* Pretoria: Unisa Press, 2003.

Lipschitz, O., and J. Blenkinsopp. *Judah and the Judeans in the Neo-Babylonian Period.* Winona Lake, Ind.: Eisenbrauns, 2003.

Lovell, A., C. Moreau, V. Yotova, F. Xiao, S. Bourgeois, D. Gehl, J. Bertran-petit, E. Schurr, and D. Labuda. "Ethiopia: Between Sub-Saharan Africa and Western Eurasia." *Annals of Human Genetics* 69 (2005): 275–287.

Lucotte, G., and P. Smets. "Origins of Falasha Jews Studied by Haplotypes of the Y Chromosome." *Human Biology* 71 (1999): 989–993.

Mathiva, M. E. R. "The BaLemba." 1972. Unpublished.

———. "The Basena/Vamwenya/Balemba." 1992. Unpublished.

———. "Lemba Characteristics." 1987. Unpublished.

———. "The Story of the Lemba People." Presentation to Zionist Lunch Club, Johannesburg, 15 October 1999. http://www.haruth.com/JewishLemba .html.

Parfitt, T. "Constructing Black Jews: Genetic Tests and the Lemba—The 'Black Jews' of South Africa." *Developing World Bioethics* 3 (2003): 112–118.

———. "Genes, Religion, and History: The Creation of a Discourse of Origin among a Judaizing African Tribe." *Jurimetrics* 42 (2002): 209–219.

———. *Journey to the Vanished City: The Search for a Lost Tribe of Israel.* New York: Vintage Books, 2000.

———. *The Lost Tribes of Israel: The History of a Myth.* London: Weidenfeld and Nicolson, 2002.

Parfitt, T., and Y. Egorova. *Genetics, Mass Media, and Identity: A Case Study of the Genetic Research on the Lemba.* Abingdon, U.K.: Routledge, 2006.

Parfitt, T., and E. Trevisan Semi. *The Beta Israel in Ethiopia and Israel: Studies on Ethiopian Jews.* Surrey: Curzon, 1999.

Ruwitah, A. "Lost Tribe, Lost Language? The Invention of a False Remba Identity." *Zimbabwea* 5 (1997): 53–71.

Sharman, J. C. *The (Pre) Proto-Bantu Expansion-Migration: Some Linguistic and Ecological Evidence.* Nairobi: University of Nairobi, Institute of African Studies, 1972.

Shen, P., T. Lavi, T. Kivisild, V. Chou, D. Sengun, D. Gefel, I. Shpirer, et al. "Reconstruction of Patrilineages and Matrilineages of Samaritans and Other Israeli Populations from Y-Chromosome and Mitochondrial DNA Sequence Variation." *Human Mutation* 24 (2004): 248–260.

Spurdle, A. B., and T. Jenkins. "The Origins of the Lemba 'Black Jews' of Southern Africa: Evidence from p12F2 and Other Y-Chromosome Markers." *American Journal of Human Genetics* 59 (1996): 1126–1133.

Thomas, M. G., T. Parfitt, D. A. Weiss, K. Skorecki, J. F. Wilson, M. le Roux, N. Bradman, and D. B. Goldstein. "Y Chromosomes Traveling South: The Cohen Modal Haplotype and the Origins of the Lemba—the 'Black Jews of Southern Africa.'" *American Journal of Human Genetics* 66 (2000): 674–686.

Thomas, M. G., K. Skorecki, H. Ben Ami, T. Parfitt, N. Bradman, and D. B. Goldstein. "Origins of Old Testament Priests." *Nature* 394 (1998): 138–140.

Zoloth, L. "Yearning for the Long Lost Home: The Lemba and Jewish Narrative of Genetic Return." *Developing World Bioethics* 3 (2003): 128–132.

Chapter 3. Looking Out for Number Two

Behar, D. M., D. Garrigan, M. E. Kaplan, Z. Mobasher, D. Rosengarten, T. M. Karafet, L. Quintana-Murci, H. Ostrer, K. Skorecki, and M. F. Hammer. "Contrasting Patterns of Y Chromosome Variation in Ashkenazi Jewish and Host Non-Jewish European Populations." *Human Genetics* 114 (2004): 354–365.

Behar, D. M., M. G. Thomas, K. Skorecki, M. F. Hammer, E. Bulygina, D. Rosengarten, A. L. Jones, et al. "Multiple Origins of Ashkenazi Levites: Y Chromosome Evidence for Both Near Eastern and European Ancestries." *American Journal of Human Genetics* 73 (2003): 768–779.

Blenkinsopp, J. *Sage, Priest, Prophet: Religious and Intellectual Leadership in Ancient Israel.* Louisville, Ky.: Westminster John Knox Press, 1995.

Bright, J. *A History of Israel.* 4th ed. Louisville, Ky.: Westminster John Knox Press, 2000.

Brook, K. A. *The Jews of Khazaria.* Northvale, N.J.: Jason Aronson, 1999.

Dimont, M. I. *Jews, God, and History.* New York: Simon and Schuster, 1962.

Geary, J., and U. Sautter. "The Sorbs, Germany." *Time Europe,* 29 August 2005.

Golden, P. B. *Nomads and Their Neighbours in the Russian Steppe: Turks, Khazars, and Qipchaqs.* Aldershot, England: Ashgate, 2003.

Koestler, A. *The Thirteenth Tribe: The Khazar Empire and Its Heritage.* London: Hutchinson, 1976.

Konner, M. *Unsettled: An Anthropology of the Jews.* New York: Viking Compass, 2003.

Nebel, A., D. Filon, M. Faerman, H. Soodyall, and A. Oppenheim. "Y Chromosome Evidence for a Founder Effect in Ashkenazi Jews." *European Journal of Human Genetics* 13 (2005): 388–391.

Nurmela, R. *The Levites: Their Emergence as a Second-Class Priesthood.* Atlanta: Scholars Press, 1998.

Potok, C. *Wanderings: Chaim Potok's History of the Jews.* New York: Knopf, 1978.

Shanks, H. *Ancient Israel: From Abraham to the Roman Destruction of the Temple.* Rev. ed. Washington, D.C.: Biblical Archaeology Society, 1999.

Skorecki, K., S. Selig, S. Blazer, R. Bradman, N. Bradman, P. J. Waburton, M. Ismajlowicz, and M. F. Hammer. "Y Chromosomes of Jewish Priests." *Nature* 385 (1997): 32.

Thomas, M. G., K. Skorecki, H. Ben Ami, T. Parfitt, N. Bradman, and D. B. Goldstein. "Origins of Old Testament Priests." *Nature* 394 (1998): 138–140.

Wexler, P. *The Ashkenazic Jews: A Slavo-Turkic People in Search of a Jewish Identity.* Columbus, Ohio: Slavica, 1993.

———. *Two-Tiered Relexification in Yiddish: Jews, Sorbs, Khazars, and the Kiev Polessian Dialect.* Berlin: Mouton de Gruyter, 2002.

Chapter 4. Those Jewish Mothers

Arnason, E. "Genetic Heterogeneity of Icelanders." *Annals of Human Genetics* 67 (2003): 5–16.

Behar, D. M., M. F. Hammer, D. Garrigan, R. Villems, B. Bonne-Tamir, M. Richards, D. Gurwitz, et al. "MtDNA Evidence for a Genetic Bottleneck in the Early History of the Ashkenazi Jewish Population." *European Journal of Human Genetics* 12 (2004): 355–364.

Behar, D. M., E. Metspalu, T. Kivisild, A. Achilli, Y. Hadid, S. Tzur, L. Pereira, et al. "The Matrilineal Ancestry of Ashkenazi Jewry: Portrait of a Recent Founder Event." *American Journal of Human Genetics* 78 (2006): 487–497.

Blady, K., and S. Kaplan. *Jewish Communities in Exotic Places*. Northvale, N.J.: Jason Aronson, 2000.

Bonne-Tamir, B., M. Korostishevsky, A. J. Redd, Y. Pel-Or, M. E. Kaplan, and M. F. Hammer. "Maternal and Paternal Lineages of the Samaritan Isolate: Mutation Rates and Time to Most Recent Common Male Ancestor." *Annals of Human Genetics* 67 (2003): 153–164.

Bradman, N., and M. Thomas. "Why Y? The Y Chromosome in the Study of Human Evolution, Migration, and Prehistory." *Science Spectra* 14 (1998): 32–37.

Bright, J. *A History of Israel*. 4th ed. Louisville, Ky.: Westminster John Knox Press, 2000.

Brook, K. A. *The Jews of Khazaria*. Northvale, N.J.: Jason Aronson, 1999.

Cahnman, W. J. *German Jewry: Its History and Sociology: Selected Essays by Werner J. Cahnman*. Ed. Joseph B. Maier, Judith Marcus, and Zoltán Tarr. New Brunswick, N.J.: Transaction Publishers, 1989.

Cavalli-Sforza, L. L., and M. W. Feldman. "The Application of Molecular Genetic Approaches to the Study of Human Evolution." *Nature Genetics* 33 (2003): S266–S275.

Chhakchhuak, L. "Bnei Menashe Rejoin the Tribe." *Jerusalem Post,* 12 October 2005.

Cohen, S. J. D. "The Matrilineal Principle in Historical Perspective." *Judaism* 34 (1985): 5–13.

Gilmore, I. "Indian 'Jews' Resist DNA Tests to Prove They Are a Lost Tribe." *Telegraph* (London), 10 November 2002.

Golden, P. B. *Nomads and Their Neighbours in the Russian Steppe: Turks, Khazars and Qipchaqs.* Aldershot, England: Ashgate, 2003.

Halkin, H. *Across the Sabbath River: In Search of a Lost Tribe of Israel.* Boston: Houghton Mifflin, 2002.

Hallgrimsson, B., B. O. Donnabhain, G. B. Walters, D. M. Cooper, D. Gudbjartsson, and K. Stefansson. "Composition of the Founding Population of Iceland: Biological Distance and Morphological Variation in Early Historic Atlantic Europe." *American Journal of Physical Anthropology* 124 (2004): 257–274.

Helgason, A., G. Nicholson, K. Stefansson, and P. Donnelly. "A Reassessment of Genetic Diversity in Icelanders: Strong Evidence from Multiple Loci for Relative Homogeneity Caused by Genetic Drift." *Annals of Human Genetics* 67 (2003): 281–297.

Helgason, A., B. Yngvadottir, B. Hrafnkelsson, J. Gulcher, and K. Stefansson. "An Icelandic Example of the Impact of Population Structure on Association Studies." *Nature Genetics* 37 (2005): 90–95.

Jobling, M. A., and C. Tyler-Smith. "The Human Y Chromosome: An Evolutionary Marker Comes of Age." *Nature Reviews Genetics* 4 (2003): 598–612.

Koestler, A. *The Thirteenth Tribe: The Khazar Empire and Its Heritage.* London: Hutchinson, 1976.

Konner, M. *Unsettled: An Anthropology of the Jews.* New York: Viking Compass, 2003.

Lucotte, G., and P. Smets. "Origins of Falasha Jews Studied by Haplotypes of the Y Chromosome." *Human Biology* 71 (1999): 989–993.

Maity, B., T. Sitalaximi, R. Trivedi, and V. K. Kashyap. "Tracking the Genetic Imprints of Lost Jewish Tribes among the Gene Pool of Kuki-Chin-Mizo Population of India." *Genome Biology* 6, no. 1 (2004), http://genomebiology.com/2004/6/1/P1.

Nebel, A., D. Filon, B. Brinkmann, P. P. Majumder, M. Faerman, and A. Oppenheim. "The Y Chromosome Pool of Jews as Part of the Genetic Landscape of the Middle East." *American Journal of Human Genetics* 69 (2001): 1095–1112.

Oota, H., W. Settheetham-Ishida, D. Tiwawech, T. Ishida, and M. Stone-king. "Human mtDNA and Y-Chromosome Variation Is Correlated with Matrilocal versus Patrilocal Residence." *Nature Genetics* 29 (2001): 20–21.

Pääbo, S., H. Poinar, D. Serre, V. Jaenicke-Despres, J. Hebler, N. Rohland, M. Kuch, J. Krause, L. Vigilant, and M. Hofreiter. "Genetic Analyses from Ancient DNA." *Annual Review of Genetics* 38 (2004): 645–679.

Poppel, S. M. *German Judaica: A Sampling of Harvard's Library Resources for the Study of German Jewry; Exhibition, May 11–21, 1982, Widener Library, Harvard University.* Cambridge, Mass.: Harvard University Library, 1982.

Seielstad, M. T., E. Minch, and L. L. Cavalli-Sforza. "Genetic Evidence for a Higher Female Migration Rate in Humans." *Nature Genetics* 20 (1998): 278–280.

Shanks, H. *Ancient Israel: From Abraham to the Roman Destruction of the Temple.* Rev. ed. Washington, D.C.: Biblical Archaeology Society, 1999.

Shen, P., T. Lavi, T. Kivisild, V. Chou, D. Sengun, D. Gefel, I. Shpirer, et al. "Reconstruction of Patrilineages and Matrilineages of Samaritans and Other Israeli Populations from Y-Chromosome and Mitochondrial DNA Sequence Variation." *Human Mutation* 24 (2004): 248–260.

Thomas, M. G., M. E. Weale, A. L. Jones, M. Richards, A. Smith, N. Red-head, A. Torroni, et al. "Founding Mothers of Jewish Communities: Geographically Separated Jewish Groups Were Independently Founded by Very Few Female Ancestors." *American Journal of Human Genetics* 70 (2002): 1411–1420.

van Straten, J. "Jewish Migrations from Germany to Poland: The Rhineland Hypothesis Revisited." *Mankind Quarterly* 44 (2004): 367–383.

Wertheimer, J. *All Quiet on the Religious Front? Jewish Unity, Denomination-alism, and Postdenominationalism in the United States.* New York: The American Jewish Committee, 2005. http://www.ajc.org/site/apps/nl/content3.asp?c=ijITI2PHKoG&b=843137&ct=1044289

Wexler, P. *The Ashkenazic Jews: A Slavo-Turkic People in Search of a Jewish Identity.* Columbus, Ohio: Slavica, 1993.

Wilder, J. A., S. B. Kingan, Z. Mobasher, M. M. Pilkington, and M. F. Hammer. "Global Patterns of Human Mitochondrial DNA and Y-

Chromosome Structure Are Not Influenced by Higher Migration Rates of Females versus Males." *Nature Genetics* 36 (2004): 1122–1125.

Chapter 5. Look on Mine Affliction

Abbott, A. "Genetic Patent Singles Out Jewish Women." *Nature* 436 (2005): 12.

Cahnman, W. J. *German Jewry: Its History and Sociology: Selected Essays by Werner J. Cahnman.* Ed. Joseph B. Maier, Judith Marcus, and Zoltán Tarr. New Brunswick, N.J.: Transaction Publishers, 1989.

Calvert, P. M., and H. Frucht. "The Genetics of Colorectal Cancer." *Annals of Internal Medicine* 137 (2002): 603–612.

Carmeli, D. B. "Prevalence of Jews as Subjects in Genetic Research: Figures, Explanation, and Potential Implications." *American Journal of Medical Genetics, Part A* 130 (2004): 76–83.

Charrow, J. "Ashkenazi Jewish Genetic Disorders." *Familial Cancer* 3 (2004): 201–206.

Cochran, G., J. Hardy, and H. Harpending. "Natural History of Ashkenazi Intelligence." *Journal of Biosocial Science* 38 (2006): 659–693.

Coughlin, S. S., M. J. Khoury, and K. K. Steinberg. "BRCA1 and BRCA2 Gene Mutations and Risk of Breast Cancer: Public Health Perspectives." *American Journal of Preventive Medicine* 16 (1999): 91–98.

Dean, M., M. Carrington, and S. J. O'Brien. "Balanced Polymorphism Selected by Genetic versus Infectious Human Disease." *Annual Review of Genomics and Human Genetics* 3 (2002): 263–292.

Eng, C. M., and R. J. Desnick. "Experiences in Molecular-Based Prenatal Screening for Ashkenazi Jewish Genetic Diseases." *Advances in Genetics* 44 (2001): 275–296.

Gason, A. A., E. Sheffield, A. Bankier, M. A. Aitken, S. Metcalfe, K. Barlow Stewart, and M. B. Delatycki. "Evaluation of a Tay-Sachs Disease Screening Program." *Clinical Genetics* 63 (2003): 386–392.

George, A. "The Rabbi's Dilemma." *New Scientist* 181 (2004): 44–47.

Germain, D. P. "Gaucher Disease: Clinical, Genetic, and Therapeutic Aspects." *Pathologie-Biologie* 52 (2004): 343–350.

Gilbert, F. "Familial Dysautonomia and the Expansion of the Ashkenazi Jewish Carrier Screening Panel." *Genetic Testing* (2001): 83–85.

Goodman, R. M. *Genetic Disorders among the Jewish People.* Baltimore: Johns Hopkins University Press, 1979.

Goodman, R. M., and A. G. Motulsky. *Genetic Diseases among Ashkenazi Jews.* New York: Raven Press, 1979.

Karpati, M., E. Gazit, B. Goldman, A. Frisch, R. Colombo, and L. Peleg. "Specific Mutations in the HEXA Gene among Iraqi Jewish Tay-Sachs Disease Carriers: Dating of Founder Ancestor." *Neurogenetics* 5 (2004): 35–40.

Kedar-Barnes, I., and P. Rozen. "The Jewish People: Their Ethnic History, Genetic Disorders, and Specific Cancer Susceptibility." *Familial Cancer* 3 (2004): 193–199.

Lynch, H. T., W. S. Rubinstein, and G. Y. Locker. "Cancer in Jews: Introduction and Overview." *Familial Cancer* 3 (2004): 177–192.

Matthijs, G. "The European Opposition against the BRCA Gene Patents." *Familial Cancer* 5 (2006): 95–102.

Motulsky, A. G. "Jewish Diseases and Origins." *Nature Genetics* 9 (1995): 99–101.

Narod, S. A., and W. D. Foulkes. "BRCA1 and BRCA2: 1994 and Beyond." *Nature Reviews Cancer* 4 (2004): 665–676.

Narod, S. A., and K. Offit. "Prevention and Management of Hereditary Breast Cancer." *Journal of Clinical Oncology* 23 (2005): 1656–1663.

Niell, B. L., J. C. Long, G. Rennert, and S. B. Gruber. "Genetic Anthropology of the Colorectal Cancer–Susceptibility Allele APC I1307K: Evidence of Genetic Drift within the Ashkenazim." *American Journal of Human Genetics* 73 (2003): 1250–1260.

Ostrer, H. "A Genetic Profile of Contemporary Jewish Populations." *Nature Reviews Genetics* 2 (2001): 891–898.

Raas-Rothschild, A., R. Bargal, S. Della Pergola, M. Zeigler, and G. Bach. "Mucolipidosis Type IV: The Origin of the Disease in the Ashkenazi Jewish Population." *European Journal of Human Genetics* 7 (1999): 496–498.

Risch, N., H. Tang, H. Katzenstein, and J. Ekstein. "Geographic Distribution of Disease Mutations in the Ashkenazi Jewish Population Supports

Genetic Drift over Selection." *American Journal of Human Genetics* 72 (2003): 812–822.

Rosen, C. "Eugenics—Sacred and Profane." *New Atlantis,* Summer 2003, 79–89.

Rosner, F. "Judaism, Genetic Screening and Genetic Therapy." *Mount Sinai Journal of Medicine* 65 (1998): 406–413.

Senior, J. "Are Jews Smarter?" *New York,* 24 October 2005.

Slatkin, M. "A Population-Genetic Test of Founder Effects and Implications for Ashkenazi Jewish Diseases." *American Journal of Human Genetics* 75 (2004): 282–293.

Zlotogora, J., and G. Bach. "The Possibility of a Selection Process in the Ashkenazi Jewish Population." *American Journal of Human Genetics* 73 (2003): 438–440.

INDEX